BASIC ANATOMY

BASIC ANATOMY
A Laboratory Manual
The Human Skeleton | The Cat

B. L. Allen
Ohio University

W. H. FREEMAN AND COMPANY
San Francisco

Printed in the United States of America.

International Standard Book Number: 0-7167-0677-6

Preface

In many elementary anatomy courses, such as those for students in physical education, pre–physical therapy, prenursing, premortuary, and medical technology, it is desirable to emphasize the structure of the human body. Seeing and handling the actual structures is very helpful for learning anatomy and understanding anatomical relationships. However, in most undergraduate colleges the use of human cadavers for dissection is not feasible, because it is difficult to obtain them and special facilities are needed for their storage. Instead the cat is often used, since it is anatomically very similar to the human, and it is less expensive and easier to store. But for study of the skeletal system, the human skeleton and the separate bones are easily obtained and stored, and the cost is not prohibitive. This laboratory manual, then, has been designed for courses in which the human skeleton and the cat are studied. Where there are pertinent differences between the cat and the human, they are noted; the student should know these differences, and make use of his own living specimen (himself) as much as possible, to help him learn and appreciate the structure of the human body, as well as that of the cat.

The material is organized systemically, but some relevant regional features are included with the study of the muscles of the appendages. The directions for the study of each system are more detailed than will be necessary for every class, and it is assumed that the instructor will omit the material that he considers nonessential for his particular course.

To aid in the identification of various structures, there are numerous illustrations of the skull and other components of the human skeleton, as well as an illustration of the articulated human skeleton and one of the cat skeleton. There are also many illustrations of the musculature and the internal organ systems of the cat. A list of definitions for special features, or markings, on bones is included for students who are studying bones in detail; it is intended as a reference, rather than as a set of terms to be memorized. Students who are particularly concerned with human muscles can draw these in place on the human skeletal diagrams provided at the end of the chapter on the muscular system.

The illustrations and the directions for study and dissection have been designed to enable the student to progress in his laboratory work with little help from the instructor. The student should always try to locate, externally on himself, as many structures as he can, and to be aware that the structures he is studying are not isolated, but in life are functionally integrated parts within a system, which in turn is functionally integrated with the other systems to form a whole working body.

I wish to express my sincere appreciation to a number of people for their help in the preparation of this manual: my students, who first thought that I should prepare a manual for them; Rush Elliott, Professor of Anatomy at Ohio University, who read the manuscript and made many valuable suggestions; Elizabeth Smith Cole, who made the original drawings of the dissected cat; Doreen Davis Masterson, who did dissection and layouts and some of the final drawings; and Edward Hanson, Julia P. Iltis, Edna Indritz, Jill Leland, and Margaret L. Muller, who also prepared final drawings.

February, 1970   B. L. Allen

Contents

Illustrations

FIGURES

HUMAN SKELETAL DIAGRAMS

Introduction to the Student: General Directions

Use of Laboratory Materials

At all times, handle with care the fragile bone specimens or other materials that may be provided for your use. When it is necessary to share specimens with others, be unselfish and courteous.

Preliminary Study

Become familiar with directional terms and planes of section that are used with reference to the body.

DIRECTIONAL TERMS

cranial, or **cephalic.** The head end. (*craniad,* or *cephalad*: toward the head.)

caudal. The tail end. (*caudad.* Toward the tail.)

ventral. The belly side. (*ventrad.* Toward the belly.)

dorsal. The back side. (*dorsad.* Toward the back.)*

superior. The upper portion, or that which is above.

inferior. The lower portion, or that which is below.

anterior. Used synonymously with "cranial" in quadrupeds, and with "ventral" in bipeds.

posterior. Used synonymously with "caudal" in quadrupeds, and with "dorsal" in bipeds.

median. The middle, or the midline.

midline. An imaginary plane that bisects the body into right and left halves.

---

*NOTE: In reference to the hindlimb, the terms "dorsal" and "ventral" differ between the cat and the human. In the cat, the anterior aspect of the hindlimb is dorsal and the posterior aspect is ventral. In the human, dorsal refers to the posterior aspect of the hindlimb, and ventral to the anterior. In this manual the terms corresponding to dorsal and ventral in the human will be used throughout to avoid confusion.

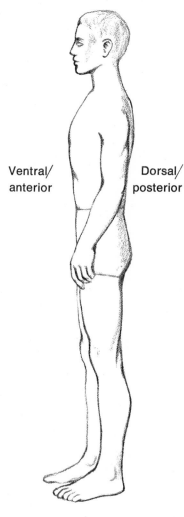

Cranial/superior

Ventral/
anterior

Dorsal/
posterior

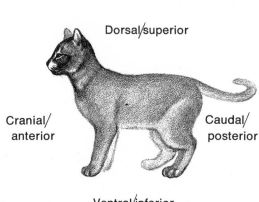

Dorsal/superior

Cranial/
anterior

Caudal/
posterior

Ventral/inferior

a

DIRECTIONAL TERMS:
(*a*) quadruped; (*b*) biped

b

Caudal/inferior

medial. Refers to a location nearer to the midline than another location. (*mediad.* Toward the midline.)

lateral. Refers to a location farther from the midline than another location. (*laterad.* Away from the midline.)

proximal. Commonly used with reference to the appendages, meaning that portion which is nearer the trunk or main body mass (nearest to point of attachment to the trunk). (*proximad.* Toward the trunk.)

distal. Used in conjunction with proximal, meaning farther away from the trunk (farthest from point of attachment). (*distad.* Away from the trunk.)

PLANES OF SECTION

transverse, or **cross,** or **horizontal.** A plane that extends from left to right and from dorsal to ventral, giving cranial and caudal portions.

longitudinal. A plane that extends from cranial to caudal; longitudinal and transverse planes intersect at right angles.

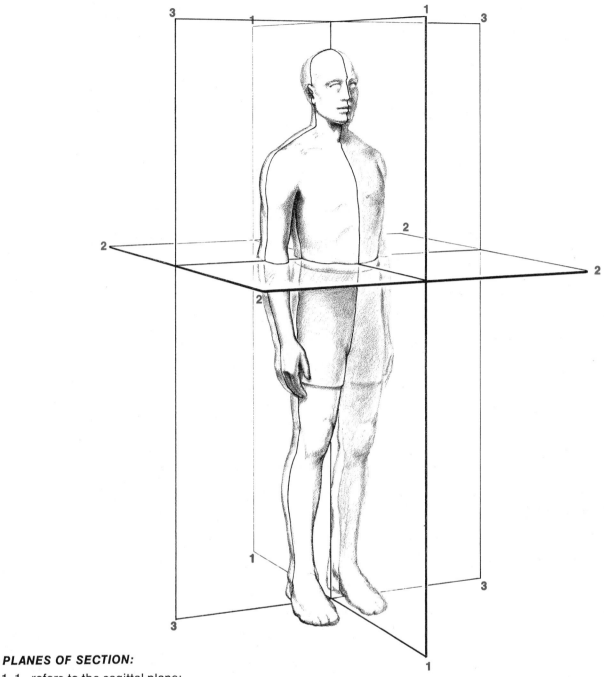

PLANES OF SECTION:

1–1   refers to the sagittal plane;
2–2,   to the transverse plane;
3–3,   to the frontal or coronal plane.

sagittal, or **midsagittal**. A longitudinal plane that passes through the midline to divide the body into right and left halves.

parasagittal. A longitudinal plane that parallels the sagittal plane to either the right or the left.

frontal, or **coronal**. A longitudinal plane that extends from left to right and cranial to caudal, giving dorsal and ventral portions.

1 SKELETAL SYSTEM OF THE HUMAN

Observe skeletons of both the human and the cat, and make comparisons. It is important to become oriented with the cat skeleton as well as with the human, since the cat will be used for dissection. Use the illustrations and descriptions that are included in this manual as study aids in identifying structures and various details indicated in the outline that follows.

The important joints should be studied in conjunction with the various bones that are involved in these joints, and also the movements that are possible in the joints.

A list of basic bone features is given on pages 2 and 4 for classes in which details of individual bones are studied.

Divisions of the Skeleton
(Figs. 1-1 and 1-2, pp. 2 and 3)

AXIAL DIVISION

Skull

Cranium. The portion of the skull that encases the brain.
Face. Region of the forehead, eyes, nose, cheeks, and jaws.
Ear ossicles. These three small bones are enclosed in the petrous portion of the temporal bone and will not be seen.
Hyoid. A single bone having no actual bony connections with the skull.

Vertebral Column. The following different regions are recognized:

Cervical. The neck.
Thoracic. The "chest."
Lumbar. The "small" of the back.
Sacral. The pelvic area.
Coccygeal, or caudal. The tail.

Thorax

Sternum. The "breastbone."
Ribs. The bones connecting the thoracic vertebrae and the sternum; all ribs together form the rib cage.

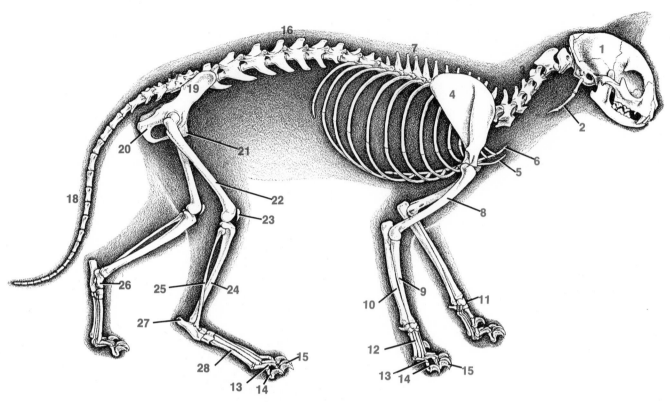

Fig. 1–1

LATERAL VIEW OF
THE CAT SKELETON

1. Skull
2. Hyoid bone
3. Cervical region
4. Scapula
5. Clavicle
6. Sternum
7. Thoracic region
8. Humerus
9. Radius
10. Ulna
11. Carpal bones
12. Metacarpal bones
13. Proximal phalanges
14. Middle phalanges
15. Distal phalanges
16. Lumbar region
17. Sacrum
18. Coccygeal region
19. Ilium
20. Ischium
21. Pubis
22. Femur
23. Patella
24. Tibia
25. Fibula
26. Talus
27. Calcaneus
28. Metatarsal bones

APPENDICULAR DIVISION

Pectoral Appendages. The cranial or superior appendages.

    Pectoral girdles. These consist of the shoulder blade and collar bone on each side.

    Pectoral extremities. The arm, forearm, and hand on each side.

Pelvic Appendages. The caudal or inferior appendages.

    Pelvic girdle. Consists of the two "hip" bones, the innominate bones, which articulate ventrally at the midline; each bone forms a joint (sacroiliac) with the sacrum.

    Pelvic extremities. The thigh, leg, and foot on each side.

Note differences between cat and human skeletons in the following:

Clavicle
Sternum
Number of vertebrae in different regions
Shape of innominate bone
Number of digits (toes) on pelvic appendage.

Basic Bone Features

Aditus. The entrance to a cavity.

Alveolus. A deep pit or socket, such as one that holds a tooth.

Condyle. A rounded or knuckle-like prominence, such as the occipital condyle; may be found at the articulation point of one bone with another.

Crest. A narrow ridge of bone.

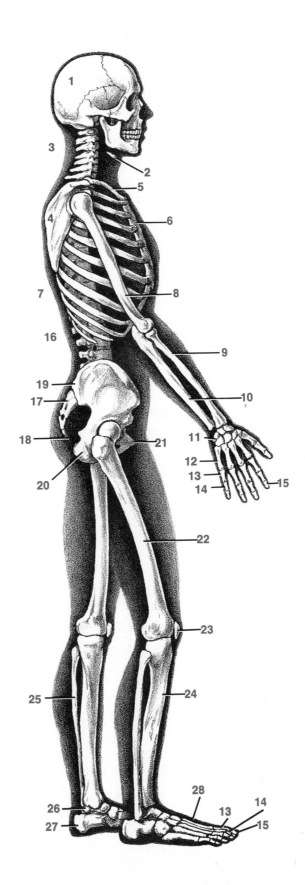

Fig. 1–2

LATERAL VIEW OF
THE HUMAN SKELETON

1. Skull
2. Hyoid bone
3. Cervical region
4. Scapula
5. Clavicle
6. Sternum
7. Thoracic region
8. Humerus
9. Radius
10. Ulna
11. Carpal bones
12. Metacarpal bones
13. Proximal phalanges
14. Middle phalanges
15. Distal phalanges
16. Lumbar region
17. Sacrum
18. Coccygeal region
19. Ilium
20. Ischium
21. Pubis
22. Femur
23. Patella
24. Tibia
25. Fibula
26. Talus
27. Calcaneus
28. Metatarsal bones

Eminence. A ridge or projection, or rounded prominence.

Epicondyle. A protuberance or prominence above a condyle.

Facet. A smooth, flattened articular surface.

Fissure. A narrow slit or cleft; often this occurs between two bones.

Foramen. An orifice for passage of blood vessels and/or nerves.

Fossa. A depression or hollow.

Head. A rounded projection beyond a constricted part or neck, as on the femur.

Line. A ridge of bone less prominent than a crest.

Meatus, or **canal.** A long, tube-like passage.

Process. Any marked bony prominence or projection.

Sinus, or **antrum.** A cavity within a bone.

Spine. A sharp, slender projection, such as the spinous process of a vertebra.

Sulcus. A furrow or groove.

Trochanter. A large process for muscle attachment, such as the greater trochanter of the femur.

Trochlea. A process shaped like a pulley.

Tubercle. A small rounded process, such as the tubercle of a rib.

Tuberosity. A large rounded process, such as the ischial tuberosity.

Details of the Axial Division of the Skeleton

Using the human bones and the illustrations provided, identify the components of the different portions of the axial division of the skeleton. Some bones are paired (2) and some are single (1). Determine the major kinds of joints that are present.

SKULL
(Figs. 1-3, 1-4, 1-5, 1-6, 1-7, 1-8, 1-9, and 1-10, pp. 5-9)

The skull contains 29 bones, including the hyoid

Cranium

Frontal—1
Parietal—2
Temporal—2
Occipital—1
Sphenoid—1
Ethmoid—1

Face

Maxillary—2
Mandible—1
Zygomatic or malar—2

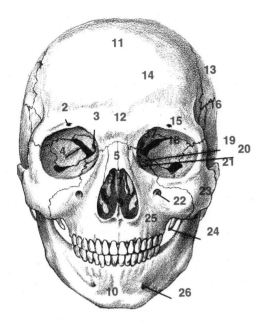

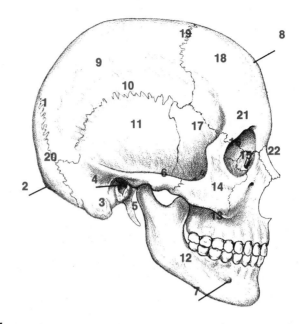

Fig. 1–3
SKULL, FRONT VIEW

1. Coronal suture
2. Supraorbital foramen
3. Optic canal
4. Superior orbital fissure
5. Nasal bone
6. Middle concha
7. Perpendicular plate of ethmoid bone
8. Inferior concha
9. Vomer
10. Mandible
11. Frontal bone
12. Glabella
13. Parietal bone
14. Eminence
15. Supraorbital margin
16. Temporal bone
17. Lesser wing of sphenoid bone
18. Greater wing of sphenoid bone
19. Ethmoid bone
20. Lacrimal bone
21. Zygomatic arch
22. Infraorbital foramen
23. Zygomatic (malar) bone
24. Styloid process
25. Maxilla
26. Mental foramen

Fig. 1–4
SKULL, SIDE VIEW

1. Lambdoidal suture
2. External occipital protuberance
3. Mastoid process
4. External auditory meatus
5. Styloid process
6. Zygomatic arch
7. Mental foramen
8. Eminence
9. Parietal bone
10. Squamous suture
11. Temporal bone

12. Mandible
13. Maxilla
14. Zygomatic bone
15. Lacrimal bone
16. Ethmoid bone
17. Greater wing of sphenoid bone
18. Frontal bone
19. Coronal suture
20. Occipital bone
21. Supraorbital margin
22. Nasal bone

Lacrimal—2
Palatine—2
Nasal—2
Vomer—1
Inferior concha or turbinate—2

Ear ossicles (will not be seen)

Malleus (hammer)—2
Incus (anvil)—2
Stapes (stirrup)—2

Hyoid—1

Note that some of the bones listed for the cranium could also be listed as face bones.

Special Features of the Skull
Sutures

Sagittal. Between the parietal bones.
Coronal. Between the frontal and parietal bones.

Fig. 1–5

SKULL OF NEWBORN
INFANT, SIDE VIEW

1. Parietal bone
2. Occipital (posterior) fontanel
3. Occipital bone
4. Mastoid (posterolateral)
   fontanel
5. Frontal (anterior) fontanel
6. Coronal suture
7. Sphenoid (anterolateral)
   fontanel
8. Frontal bone

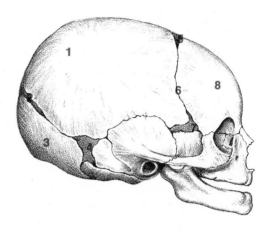

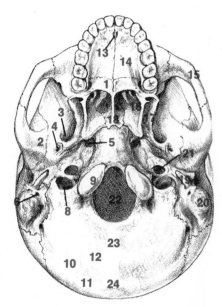

Fig. 1–6

BASE OF THE SKULL

1. Horizontal plate of palatine bone
2. Temporal bone
3. Foramen ovale
4. Foramen spinosum
5. Foramen lacerum
6. Stylomastoid foramen
7. Digastric groove
8. Jugular fossa and foramen
9. Occipital condyle
10. Occipital bone
11. Superior nuchal line
12. Inferior nuchal line
13. Incisive foramen
14. Palatine process of maxilla
15. Zygomatic arch
16. Medial pterygoid plate
    of sphenoid bone
17. Lateral pterygoid plate
    of sphenoid bone
18. Vomer
19. Styloid process
20. Mastoid process
21. Carotid canal
22. Foramen magnum
23. Median nuchal line
24. External occipital protuberance

Lambdoidal. Between the occipital and parietal bones.

Squamosal. Between the squamosal portion of the temporal bone and
   other bones.

Various others.

Junctions that mark location of major fontanels in early life

Bregma. Junction of the frontal and the two parietal bones, at the
   anterior or frontal fontanel location.

Lambda. Junction of the occipital and the two parietal bones, at the
   posterior or occipital fontanel location.

Zygomatic arch. The bony arch ventral to the ear, formed by portions
   of the temporal and zygomatic bones.

External auditory meatus. The external ear canal. Note the opening.

Foramen magnum. The large orifice in the occipital bone.

Occipital condyles. The prominences that articulate with the first cervical
   vertebra.

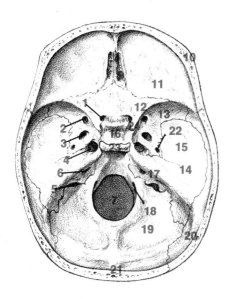

Fig. 1–7
FLOOR OF THE
CRANIAL CAVITY

1. Optic foramen
2. Foramen rotundum
3. Foramen ovale
4. Foramen lacerum
5. Jugular foramen
6. Internal auditory meatus
7. Foramen magnum
8. Crista galli
9. Cribriform plate of ethmoid bone
10. Frontal bone
11. Anterior cranial fossa
12. Lesser wing of sphenoid bone
13. Greater wing of sphenoid bone
14. Temporal
15. Middle cranial fossa
16. Sella turcica
17. Petrous portion
    of temporal bone
18. Hypoglossal canal
19. Posterior cranial fossa
20. Parietal bone
21. Occipital bone
22. Foramen spinosum
23. Posterior clinoid process
24. Anterior clinoid process

Mastoid process. The prominence located dorsal and caudal to the opening of the external auditory meatus; it contains air cells.

Foramina other than foramen magnum. Various passages for nerves and/or blood vessels.

Orbital fossae. The hollows containing the eyeballs and associated structures.

Forehead. That part of the frontal bone above the eyes and nose.

Nasal cavity. Note the septum that divides the cavity.

Cranial cavity. The space enclosed by the cranium.

Sinuses. The cavities within the frontal, maxillary, ethmoid, and sphenoid bones. They are called the **paranasal** sinuses.

Temporomandibular joint. A combination hinge and gliding joint (ginglymoarthrodial). In the living body the joint cavity is divided by a complete intra-articular disk.

Details of the Skull Bones

Identify the special features of the various bones listed below. Use the illustrations provided.

Frontal

Supraorbital margin
Supraorbital notch or foramen
Frontal sinuses (within the bone)
Frontal eminence(s)

Parietal

Parietal eminence
Temporal lines
Foramina

Temporal. (This consists of three parts: a rather thin squamous portion, a mastoid portion, and a hard petrous portion that contains the ear apparatus.)

Mastoid process
External auditory meatus
Mastoid notch, or digastric groove. Point of muscle attachment.
Zygomatic process. Part of the zygomatic arch.
Mandibular fossa. For articulation with the mandibular condyle.
Styloid process. Usually broken off; gives attachment to the hyoid bone by ligament.
Carotid canal. Gives passage to the internal carotid artery.
Jugular foramen. Gives passage to the internal jugular vein and some of the cranial nerves.
Stylomastoid foramen. Gives passage to a cranial nerve (the facial) that supplies the muscles of facial expression.
Internal auditory meatus. Gives passage to the nerves and blood vessels that supply the ear.

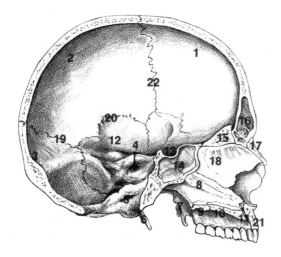

Fig. 1–8
SAGITTAL SECTION OF THE SKULL

1. Frontal bone
2. Parietal bone
3. Occipital bone
4. Internal auditory meatus
5. Hypoglossal canal
6. Styloid process
7. Pterygoid process
8. Vomer
9. Palatine bone
10. Palatine process of maxilla
11. Incisive canal
12. Temporal bone
13. Sella turcica
14. Sphenoid sinus
15. Crista galli
16. Frontal sinus
17. Nasal bone
18. Perpendicular plate of ethmoid bone
19. Lambdoidal suture
20. Squamous suture
21. Maxilla
22. Coronal suture

Occipital

Foramen magnum
Occipital condyles
Nuchal lines (superior, inferior, and median). Give attachment to muscles.
External occipital protuberance. The raised area between the superior nuchal lines; it can be palpated at the midline.
Foramina other than magnum

Sphenoid

Body. The central portion of the bone.
Greater and lesser wings. Lateral projections.
Sella turcica. The depression lodging the hypophysis.
Sphenoid sinuses. Located within the body of the bone.
Pterygoid processes. Note the plates of each process, and the fossa between the plates.
Optic foramen
Other foramina

Ethmoid

Cribriform plate. A horizontal plate with perforations for passage of olfactory nerve fibers.
Perpendicular plate. The upper part of the nasal septum.
Lateral mass. Located in the medial wall of the orbital fossa; contains air cells (ethmoid sinus).
Superior and middle conchae, or turbinates. Located in the lateral wall of the nasal cavity.

Maxillary (upper jaw)

Alveolar portion. Contains cavities for teeth.
Maxillary sinus. Located within the bone.
Palatine process. Part of the hard palate.
Infraorbital foramen. Gives passage to the infraorbital nerve and blood vessels.

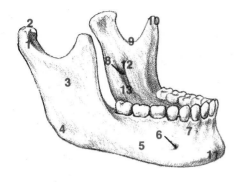

Fig. 1–9
MANDIBLE

1. Neck
2. Condyloid process
3. Ramus
4. Angle
5. Body
6. Mental foramen
7. Alveolar border
8. Mandibular foramen
9. Notch
10. Coronoid process
11. Mental symphysis
12. Lingula
13. Mylohyoid groove

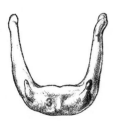

Fig. 1–10
HYOID BONE,
FRONT VIEW

1. Greater cornu
2. Lesser cornu
3. Body (basihyal)

Lacrimal groove

Zygomatic, or malar (the "cheekbone").

Lacrimal. Note the opening to the lacrimal canal, which contains the tear duct in the living body. The canal terminates in the nasal cavity.

Palatine. The horizontal plates of the two palatine bones form part of the hard palate.

Nasal. The two nasal bones form the bridge of the nose.

Vomer. Part of the lower portion of the nasal septum.

Inferior concha, or **turbinate.** Small bone below the middle concha of the ethmoid. Because these conchae are very fragile, they are often missing from the skulls in the laboratory.

Mandible (lower jaw)
Symphysis. The site at which the halves of the mandible fuse.
Body. The main portion of the bone.
Ramus
Angle
Condyle, or condyloid process. Articulates with the temporal bone.
Neck of condyle
Coronoid process. Note the mandibular notch between this process and the condyle.
Alveolar portion. Contains cavities for teeth.
Foramina

Hyoid. The hyoid is not present on a separate skull, but it should be on an articulated skeleton. It can be observed deep in the "horseshoe curve" of the mandible. It forms no joints with other bones; it is attached only by a ligament to the styloid process of the temporal bone, and is held in place by muscle attachments.

VERTEBRAL COLUMN
(Figs. 1-11, 1-12, 1-13, 1-14, 1-15, 1-16, 1-17, and 1-18, pp. 10–13)

The vertebral column in the adult contains 26 bones.

Number of Vertebrae

The number of vertebrae present, in human and cat, are listed below by region.

	Number in human	*Number in cat*
Cervical	7	7
Thoracic	12	13
Lumbar	5	7
Sacral	5 (fused to form one sacrum)	3 (fused to form one sacrum)
Coccygeal, or Caudal	2-5 (fused to form one coccyx)	many

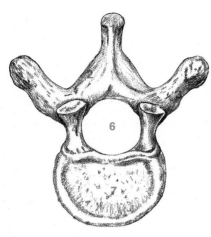

Fig. 1–11
A TYPICAL VERTEBRA

1. Spinous process
2. Lamina
3. Transverse process
4. Pedicle
5. Articular process
6. Vertebral foramen
7. Body

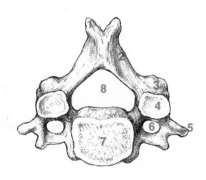

Fig. 1–12
A TYPICAL CERVICAL
VERTEBRA (FOURTH),
CRANIAL ASPECT

1. Spinous process
2. Lamina
3. Inferior articular process
4. Superior articular process
5. Transverse process
6. Transverse foramen
7. Body (centrum)
8. Vertebral foramen
9. Pedicle

Curvatures

Cervical
Thoracic
Lumbar
Sacral

Common Deformities

Scoliosis. Lateral curvature.
Lordosis. Exaggerated lumbar curvature.
Kyphosis. Exaggerated thoracic curvature.

Features of a Typical Vertebra

Body, or **centrum**

Neural arch

Dorsal spine, or **spinous process**

Lateral spines, or **transverse processes**

Vertebral foramen. (When vertebrae are fitted together, the foramina participate in the formation of a canal, the **neural canal,** that houses the spinal cord.)

Laminae. Together these support the spinous process and form the dorsal wall of the neural arch.

Pedicles. The "roots" of the neural arch.

Superior articular processes and facets

Inferior articular processes and facets

Vertebral notches. These are distinguished as "superior" and "inferior" and are located in the pedicles. (When two vertebrae are fitted together, two opposing notches form an intervertebral foramen for passage of a spinal nerve and intervertebral blood vessels.)

Articulations Between Vertebrae

On the articulated skeleton, observe the gliding joints between inferior and superior articular processes of adjacent vertebrae. Note the felt pads that simulate the **intervertebral disks.** These fibrocartilage disks are shock absorbers, and they form symphysis joints with the bodies of the vertebrae.

Special Features of Regional Vertebrae

Using the illustrations provided, identify the features of a typical vertebra from each region of the vertebral column.

Cervical Region. Note the short, bifid spinous process of a typical cervical vertebra. Because C-7 has a spine resembling those of the thoracic vertebrae and can be palpated projecting out at the base of the neck, it is called the **vertebra prominens.** Note the **transverse foramen** on each side. These foramina, which are distinctive to the cervical vertebrae, give passage to the vertebral artery and vein.

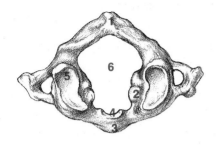

Fig. 1–13
FIRST CERVICAL VERTEBRA,
CRANIAL ASPECT

1. Posterior tubercle
2. Tubercle for transverse ligament
3. Anterior tubercle
4. Facet for odontoid process of axis
5. Superior articular facet
6. Vertebral foramen

The first two cervical vertebrae are different from the others. The first one is called the **atlas**; and the second, the **axis**, or **epistropheus.** These two have a special articulation between them that is not present in the others. The first also has special articular surfaces for the occipital condyles. Find these articular surfaces, and note also that the centrum and spinous process are greatly underdeveloped. Find the articular surface for the **dens**, or **odontoid process,** of the axis. On C-2 find the dens, as well as other features typical of a cervical vertebra. The articulations between the atlas and the occipital bone are called the **atlantoöccipital joints.** These are condyloid joints. The articulation between the atlas and the dens of the axis is called the **atlantoaxial,** or **atlantoepistropheal, joint.** This is a pivot, or trochoid joint.

Thoracic Region. Note the long spinous process of a typical thoracic vertebra. Note articular facets for ribs on the body and the transverse processes. These articular surfaces for ribs distinguish a vertebra of the thoracic region. The long, sharp spine is also a diagnostic feature of all but the last one or two vertebrae, which have a spine resembling those of the lumbar region.

Typically, the head of a rib articulates with demifacets (half facets) on the bodies of two adjacent vertebrae, and the tubercle of the rib articulates with the facet on the transverse process of the caudally adjacent vertebra.

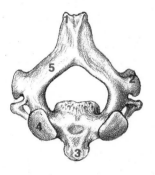

Fig. 1–14
SECOND CERVICAL VERTEBRA,
CRANIAL ASPECT

1. Spinous process
2. Inferior articular process
3. Odontoid process (dens)
4. Superior articular facet
5. Lamina

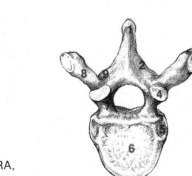

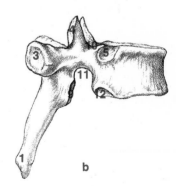

a   b

Fig. 1–15
THORACIC VERTEBRA:
(*a*) cranial aspect;
(*b*) lateral aspect

1. Spinous process
2. Lamina
3. Facet for tubercle of rib
4. Superior articular process
5. Superior demifacet
   for head of rib
6. Body (centrum)
7. Pedicle
8. Transverse process
9. Inferior articular process
10. Superior notch
11. Inferior notch
12. Inferior demifacet for head of rib

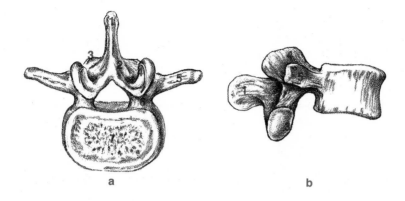

a   b

Fig. 1–16
LUMBAR VERTEBRA:
(a) cranial aspect;
(b) lateral aspect

1. Spinous process
2. Lamina
3. Inferior articular process
4. Superior articular process
5. Transverse process
6. Pedicle
7. Body (centrum)

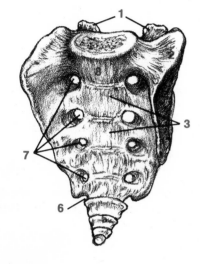

Fig. 1–17
SACRUM AND COCCYX,
VENTROLATERAL VIEW

1. Superior articular processes
2. Lateral mass
3. Body
4. Articular surface
5. Coccyx
6. Apex of sacrum
7. Anterior sacral foramina
8. Promontory

Lumbar Region. Note the broad, blunt spinous process and the comparatively slender transverse processes of a lumbar vertebra. The bodies of the lumbar vertebrae are larger than those of the thoracic vertebrae. (The bodies of the vertebrae are increasingly larger along the column in the caudal direction up to the sacral region, where the vertebrae fuse.)

Sacral Region. Five vertebrae have fused to form the sacrum, which provides the dorsal wall of the pelvic cavity. Find the features listed:

Body
Promontory
Alae or lateral masses
Sacral foramina: anterior and posterior
Superior articular processes and facets
Articular surfaces for the ilium
Lateral crests. These are lateral to the posterior foramina, and represent transverse processes.
Median crest. The crest at the median line that represents spinous processes.

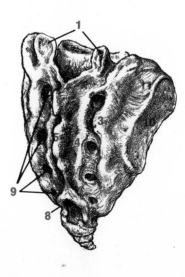

Fig. 1–18
SACRUM AND COCCYX,
DORSOLATERAL VIEW

1. Superior articular processes
2. Neural canal
3. Lateral sacral crest
4. Intermediate sacral crest
5. Median sacral crest
6. Hiatus
7. Coccyx
8. Apex of sacrum
9. Posterior sacral foramina
10. Articular surface
11. Lateral mass

Apex
Sacral hiatus. Located on the dorsal side.
Sacral portion of the neural canal
Differences in the sacrum of male and female. Note that the sacrum in the
male is narrow and more curved; that in the female is broad and
less curved.

THORAX
(Figs. 1-19 and 1-20, pp. 13 and 14)

The thorax contains 25 bones (in addition to 12 thoracic vertebrae) in
the human.

Note that the thorax is somewhat cone-shaped, being narrower at the
cranial end, the apex, than at the caudal end, the base. In addition to the
thoracic vertebrae, the bones of the thorax are the sternum (1) and the
ribs (12 pairs, or 24). The bones form a bony cage.

Sternum

Identify the parts:
Manubrium
Body, or gladiolus
Xiphoid process

Observe the following features:
Suprasternal, or jugular notch. This indentation can be palpated at
midline, at the cranial extremity of the sternum.
Clavicular notch. The site at which the clavicle articulates.
Costal notches. The sites at which the ribs articulate.
Sternal angle. A slight ridge that can be palpated at the junction of
the manubrium and the body; this angle marks the level at which
the costal cartilage of the second rib joins the sternum.

Note the articulations of the sternum:
Sternoclavicular. A gliding joint.
Sternocostal. All are gliding joints, except the first, which is a
synchondrosis.

Ribs

Identify the parts:
Body, or shaft
Head
Neck
Tubercle
Costal cartilage

Observe the features:
Angle
Articular facets on the head and tubercle

Note the articulations of the ribs:
Sternocostal. Noted above.
Costovertebral. All are gliding joints.

Ribs 1 through 7 are called **true,** or **sternal** ribs because they articulate
directly with the sternum via their costal cartilages. Note that the first

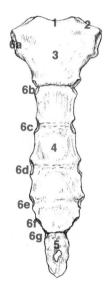

Fig. 1–19
STERNUM,
VENTRAL SURFACE

1. Jugular notch
2. Clavicular notch
3. Manubrium
4. Body
5. Xiphoid process
6. Costal notches
   a. First costal notch
   b. Second costal notch
   c. Third costal notch
   d. Fourth costal notch
   e. Fifth costal notch
   f. Sixth costal notch
   g. Seventh costal notch

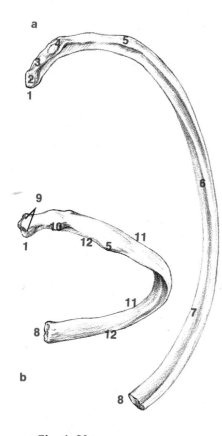

Fig. 1–20
A TYPICAL RIB (RIGHT):
(a) inferior surface;
(b) dorsal view

1. Dorsal (vertebral) extremity
2. Head
3. Neck
4. Tubercle
5. Angle
6. Costal groove
7. Body
8. Ventral extremity
9. Articular facets for bodies of vertebrae
10. Articular facet for a transverse process
11. Cranial border
12. Caudal border

rib is much shorter, broader, and more curved than the others. Ribs 8, 9 and 10 are called **false**, or **asternal** ribs because they do not articulate directly with the sternum. The costal cartilage of each articulates with the costal cartilage of the rib just cranial to it. Ribs 11 and 12 are also false, or asternal ribs, but they are called **floating** ribs because their ventral ends are free.

Note the curved downward slope of the ribs from the vertebrae to the ventral side of the body, and the upward (cranial) swing of the costal cartilage. A transverse section through the thorax would cut across more than one rib.

(A thirteenth rib, known as a gorilla rib, is occasionally present in the human. Its name is derived from the fact that gorillas usually have 13 pairs of ribs. Cats, too, typically have 13 pairs of ribs.)

Details of the Appendicular Division of the Skeleton

Proceed with this study as you did for the axial division of the skeleton.

PECTORAL APPENDAGES
(Figs. 1-21, 1-22, 1-23, 1-24, 1-25, 1-26, 1-27, and 1-28, pp.15–18)

In the adult, the pectoral appendages contain 64 bones.

Pectoral Girdles. These attach the bones of the pectoral extremities to the axial division of the skeleton.

Scapula (shoulder blade)—1 on each side.
Clavicle (collar bone)—1 on each side.

Pectoral Extremities

Arms
Humerus—1 in each arm.
Forearms
Radius (on thumb side)—1 in each forearm.
Ulna (on little-finger side)—1 in each forearm.
Hands
Carpals (wrist bones)—8 in each hand.
Metacarpals—5 in each hand.
Phalanges (bones of the digits)—2 in each thumb, 3 in each finger.

Details of Bones of the Pectoral Appendage

Scapula. A large, flat, triangular bone.

Borders
Superior, or cranial
Lateral, or axillary
Medial, or vertebral
Angles
Superior, or cranial

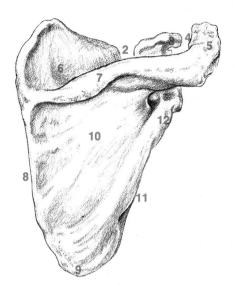

Fig. 1–22
SCAPULA (RIGHT),
VENTRAL ASPECT

1. Acromion
2. Facet for clavicle
3. Coracoid process
4. Superior border
5. Superior angle
6. Glenoid cavity
7. Lateral border
8. Inferior angle
9. Medial border
10. Subscapular fossa

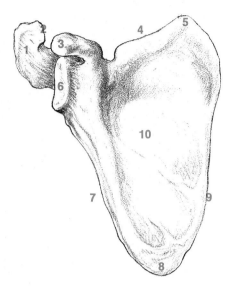

Fig. 1–21
SCAPULA (RIGHT),
DORSAL ASPECT

1. Superior angle
2. Suprascapular notch
3. Coracoid process
4. Facet for clavicle
5. Acromion
6. Supraspinous fossa
7. Spine
8. Medial border
9. Inferior angle
10. Infraspinous fossa
11. Lateral border
12. Infraglenoid tubercle

Inferior, or caudal
Lateral, or acromial
Fossae
Supraspinous. Dorsal; craniad of the spine.
Infraspinous. Dorsal; caudad of the spine.
Subscapular. Ventral.
Glenoid. Faces laterally; articulates with the head of the humerus.
Other features
Acromion
Spine
Coracoid process
Scapular notch. Located in the superior border.
Infraglenoid tubercle

Clavicle. A long bone with a double curvature. The ventral border is concave on the lateral third; the medial two-thirds is curved so that it is convex ventrally and concave dorsally.

Sternal end. Triangular, with the apex directed downward; it articulates with the manubrium of the sternum.
Acromial end. This is more flat than the sternal end; it articulates with the acromion of the scapula.
Coracoid, or conoid tubercle. Located near the acromial end, dorsal and somewhat caudal; it is attached by a ligament to the coracoid process of the scapula.
Deltoid area. A rough area at the ventral border of the acromial end; part of the deltoid muscle arises here.

Humerus

Head. Located on the medial side.
Greater tubercle, or tuberosity.
Lesser tubercle, or tuberosity.
Anatomical neck. A slight constriction between the head and the tubercles and shaft.
Surgical neck. A constriction of the shaft below the head and tubercles.

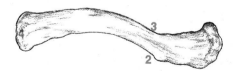

Fig. 1–23
CLAVICLE (RIGHT),
CRANIAL ASPECT

1. Acromial extremity
2. Conoid tubercle
3. Deltoid tubercle
4. Sternal extremity

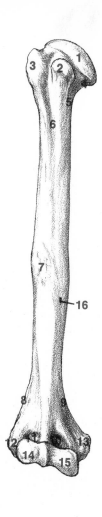

Fig. 1–24
HUMERUS (RIGHT),
VENTRAL SURFACE

1. Head
2. Lesser tubercle
3. Greater tubercle
4. Anatomical neck
5. Surgical neck
6. Intertubercular groove
7. Deltoid tuberosity
8. Lateral supracondylar ridge
9. Medial supracondylar ridge
10. Coronoid fossa
11. Radial fossa
12. Lateral epicondyle
13. Medial epicondyle
14. Capitulum
15. Trochlea
16. Nutrient foramen

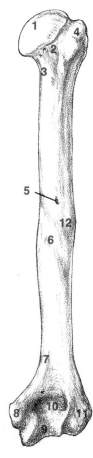

Fig. 1–25
HUMERUS (RIGHT),
DORSAL SURFACE

1. Head
2. Anatomical neck
3. Surgical neck
4. Greater tubercle
5. Nutrient foramen
6. Groove for radial nerve
7. Medial supracondylar ridge
8. Medial epicondyle
9. Trochlea
10. Olecranon fossa
11. Lateral epicondyle
12. Deltoid tuberosity

Intertubercular, or bicipital groove. For the passage of a tendon of the biceps brachii muscle.

Deltoid tuberosity. An elevation on the ventrolateral surface. This is the site of insertion for the deltoid muscle.

Spiral, or musculospiral, or radial groove. A shallow, oblique depression crossed by the radial nerve.

Medial and lateral epicondyles

Medial and lateral supracondylar ridges

Lateral condyle, or capitulum. Articulates with the head of the radius.

Medial condyle, or trochlea. Articulates with the semilunar notch of the ulna.

Olecranon fossa. Located dorsally at the distal end; it receives an ulnar process, the olecranon, when the forearm is extended.

Coronoid fossa. Located ventrally at the distal end, superior to the trochlea; it receives the coronoid process of the ulna when the forearm is flexed.

Radial fossa. Located ventrally at the distal end, superior to the capitulum; it receives the ventral portion of the head of the radius when the forearm is flexed.

Always note, in the bones, the nutrient foramina that blood vessels pass through.

Ulna. The medial bone of the forearm; it has a triangular shape in cross section.

Olecranon process (elbow bone). Located dorsally at the proximal end.

Fig. 1–26
ULNA AND RADIUS (RIGHT),
VENTRAL SURFACES

1. Ulna
2. Radius
3. Radial notch of ulna
4. Olecranon process
5. Semilunar notch
6. Coronoid process
7. Head of radius
8. **Ulnar tuberosity**
9. Supinator crest
10. Radial tuberosity
11. Nutrient foramina
12. Interosseous margins
13. Styloid process of radius
14. Styloid process of ulna
15. Head of ulna

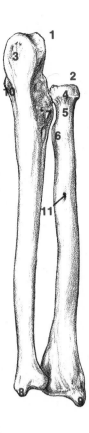

Fig. 1–27
ULNA AND RADIUS (RIGHT),
DORSAL SURFACES

1. Ulna
2. Radius
3. Olecranon
4. Articular circumference of head of radius
5. Neck of radius
6. Radial tuberosity
7. Supinator crest
8. Styloid process of ulna
9. Styloid process of radius
10. Coronoid process
11. Nutrient foramen

Coronoid process. Located ventrally at the proximal end.

Semilunar, or trochlear notch. Articulates with the humerus.

Radial notch. On the lateral side of the proximal end; it articulates with the radius.

Head. Located at the distal end; it has an articular surface for the radius.

Styloid process. Located at the distal end.

Interosseous margin or crest. Located on the lateral side.

Radius. The lateral bone of the forearm.

Head. Located at the proximal end. The shallow depression articulates with the lateral condyle, or capitulum, of the humerus; the circumference of the head articulates with the radial notch of the ulna.

Radial, or bicipital tuberosity. A protuberance for insertion of the biceps brachii muscle.

Styloid process. Located at the distal end, on the lateral side.

Ulnar notch. Located at the distal end, on the medial side; it articulates with the head of the ulna.

Interosseous margin or crest. Located on the medial side.

Observe the distal end of the radius, noting the smooth flattened ventral surface and the ridged and grooved dorsal surface, and the articular surfaces for carpal bones.

Carpals. These are listed from thumb side to little finger side for each row.

Proximal row
Scaphoid, or navicular
Lunate

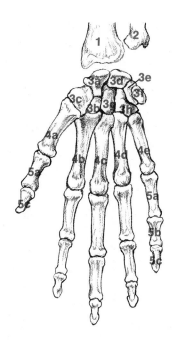

Fig. 1-28
BONES OF THE HAND
(RIGHT), PALMAR SURFACE

1. Radius
2. Ulna
3. Carpals
    a. Navicular (scaphoid)
    b. Trapezoideum
       (lesser multangular)
    c. Trapezium
       (greater multangular)
    d. Lunate (semilunar)
    e. Triquetrum (triangular)
    f. Pisiform
    g. Capitate
    h. Hamate (unciform)
4. Metacarpals
    a. First metacarpal
    b. Second metacarpal
    c. Third metacarpal
    d. Fourth metacarpal
    e. Fifth metacarpal
5. Phalanges
    a. Proximal phalanx
    b. Middle phalanx
    c. Distal phalanx

Triangular, or triquetrum
Pisiform. Small, rounded elevation on the little finger side.
Distal row
Trapezium, or greater multangular
Trapezoideum, or lesser multangular
Capitate
Hamate

Metacarpals. These are not named, but only numbered as follows, beginning with the thumb side: 1st, 2nd, 3rd, 4th, 5th.

Phalanges. (Singular term: phalanx)

Two in each thumb: proximal, distal.
Three in each finger: proximal, middle, distal.

Articulations of the Pectoral Appendage

Sternoclavicular joint. The gliding joint between the sternum and the clavicle.

Acromioclavicular joint. The gliding joint between the acromion of the scapula and the clavicle.

Shoulder joint (glenohumeral or scapulohumeral). A ball-and-socket joint between the glenoid fossa of the scapula and the head of the humerus.

Elbow joint. Between the medial and lateral condyles of the humerus, and the proximal ends of the ulna and radius (at the semi-lunar notch of the ulna and the head of the radius). Functionally, this is a hinge joint, but anatomically the radiohumeral component is a gliding joint.

Radioulnar joints

Proximal. That between the radial notch of the ulna, and the circumference of the head of the radius. It is a pivot, or trochoid joint.
Middle. A fibrous interosseous membrane connects the shafts of the radius and ulna. This joint may be classified as a syndesmosis.
Distal. Between the ulnar notch of the radius and the head of the ulna; a pivot, or trochoid joint.

Wrist joint. Between the proximal carpals (except for the pisiform), and the distal end of the radius and a cartilage disk that extends from the radius over the distal end of the ulna; this is a condyloid joint.

Intercarpal joints. Those between the carpals; all are gliding joints.

Carpometacarpal joints. Between the carpals and the metacarpals. All are gliding joints except the one between the trapezium (greater multangular) and the first metacarpal, which is a saddle joint.

Intermetacarpal joints. Between the metacarpals; these are gliding joints.

Metacarpophalangeal joints. Between the metacarpals and the proximal phalanges; generally classified as condyloid joints, but sometimes as ball-and-socket.

Interphalangeal joints. Between the phalanges; all are hinge joints.

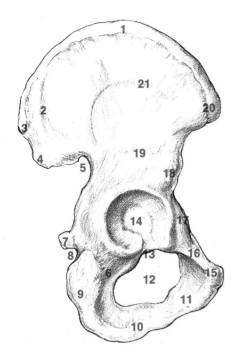

Fig. 1–29
INNOMINATE BONE (RIGHT),
EXTERNAL ASPECT

1. Crest of ilium
2. Posterior gluteal line
3. Posterior superior iliac spine
4. Posterior inferior iliac spine
5. Greater sciatic notch
6. Superior ramus of ischium
7. Ischial spine
8. Lesser sciatic notch
9. Ischial tuberosity
10. Inferior ramus of ischium
11. Inferior ramus of pubis
12. Obturator foramen
13. Acetabular notch
14. Acetabulum
15. Pubic tubercle
16. Superior ramus of pubis
17. Iliopectineal eminence
18. Anterior inferior iliac spine
19. Inferior gluteal line
20. Anterior superior iliac spine
21. Anterior gluteal line

PELVIC APPENDAGES

(Figs. 1-29, 1-30, 1-31, 1-32, 1-33, 1-34, 1-35, 1-36, 1-37, 1-38, and 1-39, pp. 19–24)

In the adult, the pelvic appendages contain 62 bones.

Pelvic girdle. Attaches the bones of the pelvic extremities to the axial division of the skeleton.

> Innominate bones—2. Each innominate bone is composed of three parts that are separately developed, but fused.
> Ilium. The cranial, flared part.
> Ischium. The caudal, dorsal part.
> Pubis. The ventral part.

Pelvic extremities

Thighs
> Femur—1 in each thigh.

Legs
> Tibia (the large bone on the medial side)—1 in each leg.
> Fibula (the small lateral bone)—1 in each leg.

Feet
> Tarsals—7 in each foot.
> Metatarsals—5 in each foot.
> Phalanges (bones of the digits)—2 in each large toe, and 3 in each of the other toes.

Kneecap
> Patella—1 in each extremity.

Details of Bones of the Pelvic Appendage

Innominate bones

Ilium
> Iliac crest
> Iliac spines. Distinguished as the anterior superior, anterior inferior, posterior superior, and posterior inferior.
> Iliac fossa. Located on the inner surface.
> Gluteal lines. Distinguished as the posterior, anterior or middle, and inferior; all are located on the outer surface.
> Greater sciatic notch
> Articular surface for the sacrum

Ischium
> Ischial spine
> Ischial tuberosity
> Lesser sciatic notch
> Superior and inferior rami

Pubis
> Superior and inferior rami
> Pubic arch. The arch formed by the articulated pubic bones.
> Pubic symphysis. The joint between the pubic bones.

Other features
> Acetabulum. Note the lunate articular surface for articulation with the head of the femur.

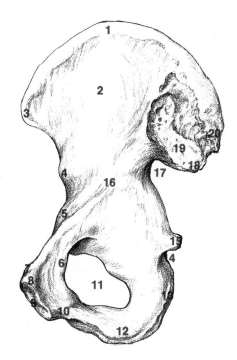

Fig. 1–30
INNOMINATE BONE (RIGHT),
INTERNAL ASPECT

1. Crest of ilium
2. Iliac fossa
3. Anterior superior iliac spine
4. Anterior inferior iliac spine
5. Iliopectineal eminence
6. Superior ramus of pubis
7. Pubic tubercle
8. Pubic crest
9. Pubic symphysis (articular surface)
10. Inferior ramus of pubis
11. Obturator foramen
12. Inferior ramus of ischium
13. Ischial tuberosity
14. Lesser sciatic notch
15. Ischial spine
16. Arcuate line
17. Greater sciatic notch
18. Posterior inferior iliac spine
19. Articular surface for sacrum
20. Posterior superior iliac spine

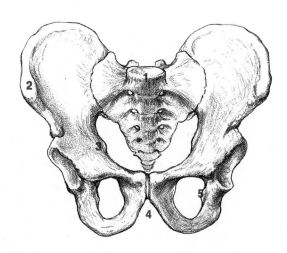

Fig. 1–31
PELVIS (MALE)

1. Sacral promontory
2. Anterior superior iliac spine
3. Arcuate line
4. Pubic arch
5. Ischial spine

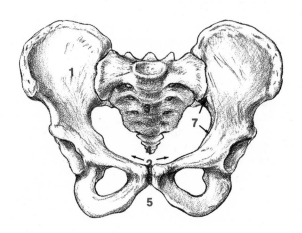

Fig. 1–32
PELVIS (FEMALE)

1. Greater (false) pelvis
2. Lesser (true) pelvis
3. Sacrum
4. Coccyx
5. Pubic arch
6. Pubic symphysis
7. Brim of lesser pelvis

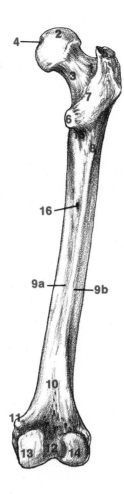

Fig. 1–34
FEMUR (RIGHT),
DORSAL SURFACE

1. Trochanteric fossa
2. Head
3. Neck
4. Fovea capitis
5. Greater trochanter
6. Lesser trochanter
7. Intertrochanteric crest
8. Gluteal tuberosity
9. Linea aspera
    a. Medial lip
    b. Lateral lip
10. Popliteal surface
11. Adductor tubercle
12. Intercondylar fossa (notch)
13. Medial condyle
14. Lateral condyle
15. Pectineal line
16. Nutrient foramen

Fig. 1–33
FEMUR (RIGHT),
VENTRAL SURFACE

1. Head
2. Neck
3. Greater trochanter
4. Intertrochanteric line
5. Lesser trochanter
6. Patellar surface
7. Adductor tubercle
8. Lateral epicondyle
9. Lateral condyle
10. Medial epicondyle
11. Medial condyle
12. Shaft

Obturator foramen

On an articulated pelvis, note the following:

Pelvic brim, or inlet. The inlet to the true pelvis.

Pelvic outlet. Caudal to the brim or inlet.

True (lesser) pelvis. Between inlet and outlet.

False (greater) pelvis. Cranial to the brim or inlet of the true pelvis.

Note also the differences between male and female in their articulated pelves. In most males the angle formed by the pubic arch is less than a right angle; in the female it is a right angle, or greater, so that the ischial tuberosities are turned outward more. In the female the ilium is flared more laterally, and the pelvic inlet is proportionately broader from side to side (left to right) than that of the male, so that it is more circular. The female sacrum is shorter, broader, and less curved than that of the male.

Femur

Head. Located on the medial side of the proximal end.

Fovea capitis. A depression in the head for ligament attachment.

Neck

Greater trochanter. The prominence on the lateral side.

Lesser trochanter. Located on the dorsomedial side.

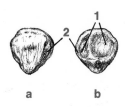

Fig. 1–35
PATELLA (RIGHT):
(a) ventral surface;
(b) dorsal surface

1. Articular surfaces
   for femur
2. Medial border
3. Point of attachment
   of patellar ligament

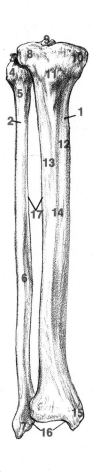

Fig. 1–36
TIBIA AND FIBULA (RIGHT),
VENTRAL SURFACES

1. Tibia
2. Fibula
3. Styloid process of fibula
4. Head of fibula
5. Neck of fibula
6. Anterior crest of fibula
7. Lateral maleolus
8. Lateral condyle of tibia
9. Intercondylar eminence
10. Medial condyle of tibia
11. Tibial tuberosity
12. Medial surface of tibia
13. Lateral surface of tibia
14. Anterior crest of tibia
15. Medial maleolus
16. Articular surfaces for talus
17. Interosseous borders

Trochanteric fossa. Located on the dorsal side, medial to the greater
trochanter.
Intertrochanteric line. On the ventral surface.
Intertrochanteric crest. On the dorsal surface.
Linea aspera. On the dorsal surface.
Medial and lateral condyles. These articulate with the tibia.
Medial and lateral epicondyles
Adductor tubercle. On the medial side at the distal end.
Intercondylar notch, or fossa. The depression between the condyles.
Trochlea (patellar surface)
Popliteal surface. The dorsal surface, at the distal end.

Patella. The kneecap; it is a sesamoid bone.

Tibia (shin bone)

Medial and lateral condyles. These articulate with the condyles of the
femur.
Intercondylar eminence
Tibial tuberosity. A rounded process on the ventral surface.
Anterior, or tibial crest. Located on the ventral surface.
Popliteal, or soleal line. On the dorsal surface.
Medial malleolus. The projection at the distal end.
Articular surfaces for the fibula and the talus (a tarsal bone). There
are two articular surfaces for the fibula, one at the proximal end
and one at the distal end; the articular surface for the talus is at the
distal end.
Interosseous border or crest. Located on the lateral side.

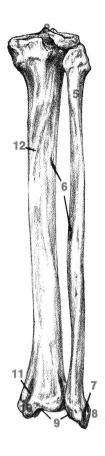

Fig. 1–37
TIBIA AND FIBULA (RIGHT), DORSAL SURFACES

1. Medial condyle of tibia
2. Intercondylar eminence
3. Lateral condyle of tibia
4. Head of fibula
5. Neck of fibula
6. Nutrient foramen
7. Groove for tendons of peroneus longus and peroneus brevis
8. Lateral maleolus
9. Articular surfaces for talus
10. Medial maleolus
11. Groove for tendons of tibialis posterior and flexor digitorum longus
12. Soleal (popliteal) line

Fibula

Lateral malleolus. Located at the distal end.

Articular surfaces. One for the tibia at the proximal end, and one for the tibia and one for the talus at the distal end.

Interosseous border or crest. On the medial side.

Tarsals

Calcaneus. The heel bone.

Talus, or astragalus

Navicular. On the medial side.

Cuboid. On the lateral side.

Cuneiforms
  medial, or I
  intermediate, or II
  lateral, or III

Metatarsals. These are comparable to the metacarpals of the hand. Like the metacarpals, they have no names, and are numbered as follows, beginning on the medial (or large toe) side: 1st, 2nd, 3rd, 4th, 5th.

Phalanges. These are the bones of the digits, comparable to the phalanges of the fingers.
  Two in each large toe: proximal, distal.
  Three in each smaller toe: proximal, middle, distal.

Arches of the Foot

Transverse
Longitudinal
  The two-way arch construction makes a highly stable base.

Articulations of the Pelvic Appendage

Sacroiliac joint. That between the sacrum and the ilium; a synchondrosis in the adult.

Interpubic joint. That between the two pubic bones; this is known as the **pubic symphysis.**

Hip joint. The joint between the acetabulum of the innominate bone and the head of the femur; a ball-and-socket joint.

Knee joint. That between the femur and the tibia, and between the

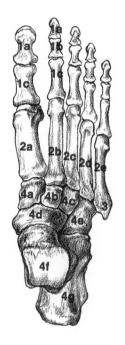

Fig. 1–39
ARCHES OF THE FOOT
(RIGHT), MEDIAL ASPECT

1. Distal phalanx
2. Proximal phalanx
3. Head of first metatarsal
4. First metatarsal
5. Fifth metatarsal
6. Medial (first) cuneiform
7. Intermediate (second) cuneiform
8. Navicular (scaphoid)
9. Cuboid
10. Talus (astragalus)
11. Calcaneus
12. Medial longitudinal arch
13. Lateral longitudinal arch
14. Transverse arch

Fig. 1–38
BONES OF THE FOOT
(RIGHT), UPPER SURFACE

1. Phalanges
   a. Distal phalanx
   b. Middle phalanx
   c. Proximal phalanx
2. Metatarsals
   a. First metatarsal
   b. Second metatarsal
   c. Third metatarsal
   d. Fourth metatarsal
   e. Fifth metatarsal
3. Tuberosity of fifth metatarsal
4. Tarsals
   a. Medial (first) cuneiform
   b. Intermediate (second) cuneiform
   c. Lateral (third) cuneiform
   d. Navicular (scaphoid)
   e. Cuboid
   f. Talus (astragalus)
   g. Calcaneus

femur and the patella. Functionally this is a hinge joint; anatomically it is made up of three joints: two condyloid joints between the condyles of the femur and the tibia, and a gliding joint between the femur and the patella.

Tibiofibular joints
Proximal. A gliding joint.
Middle. A fibrous interosseous membrane connects the shafts of the tibia and fibula; this may be classified as a syndesmosis.
Distal. A syndesmosis.

Ankle joint (talocrural). That between the talus and tibia and fibula; it is a hinge joint.

Intertarsal joints. Those between the tarsal bones; all are gliding joints. (Eversion and inversion movements of the foot are produced in the intertarsal joints.)

Tarsometatarsal joints. Those between the tarsal and metatarsal bones; all are gliding joints.

Intermetatarsal joints. Those between the metatarsals; all are gliding joints.

Metatarsophalangeal joints. Those between the metatarsals and the proximal phalanges. Generally classified as condyloid joints, but sometimes as ball-and-socket.

Interphalangeal joints. Those between the phalanges; all are hinge joints.

Materials

1½ yards plastic for each cat
1½ yards *un*bleached muslin for each cat
1 dissecting kit for each student

Removing the Skin

The embalmed cat usually comes in a plastic bag, since it is obtained from a biological supply house. If the bag has not been opened and the excess fluid drained from it, do this; then slit the bag along one side and across the closed end. The plastic will be a table cover to work on.

Place the cat dorsal side up, and grasp the skin at the nape of the neck. Using the scalpel or scissors, make an incision across the nape of the neck just barely through the skin; then make through the skin a midline incision that extends forward over the skull, down the back, and about two inches onto the tail. Sever the tail at the point where the incision ends, using bone shears or saw, and discard it.

Carefully remove the skin, working from the center of the back toward the sides. Pull the skin away from the body, using your fingers as much as possible. Where it is necessary to use the scalpel, keep the sharp edge directed toward the skin and away from underlying muscles. As you peel the skin off, note the nerves and blood vessels coursing in segmental arrangement under the skin.

It will be necessary to make other incisions as you proceed in the appendage areas. When you have removed the skin as far to the sides as possible, proceed to the ventral side. You may find it more convenient to make another incision along the midventral line and again dissect out toward the sides. When removing the skin from the head and face, leave the superficial nerves and blood vessels as intact as possible.

In the trunk area an extensive muscle will be removed with the skin. This muscle, which is closely bound to the underside of the skin, is the **cutaneous maximus**, and it is not present in the human. Fibers of the cutaneous maximus merge with the **latissimus dorsi** and **pectoralis** muscles (Fig. 3-4, p. 32) in the axillary region, and you must take care that parts of these latter muscles are not removed. In the neck region another cutaneous muscle, the **platysma,** will probably be removed with the skin.

Try to leave intact the cephalic vein (Fig. 3-1, p. 28) on the dorsal side of the pectoral appendage and the greater saphenous vein (Fig. 3-15, p. 54) on the medial side of the pelvic appendage. If it is possible, also leave intact the lesser saphenous vein (Fig. 3-1) on the dorsal side of the pelvic appendage.

The female cat will have varying amounts of mammary gland tissue on the ventral side of the thorax and abdomen. This should be removed, but do not damage the underlying muscles. If you are dissecting a male cat, be careful that reproductive structures in the inguinal and scrotal regions (Fig. 5-2, p. 108) are not destroyed. Do *not* remove the scrotal integument.

Remove the skin from the appendages to approximately the wrist and ankle joints; the remaining skin can be removed when the muscles of the region are dissected.

Clean off as much fat and other loose subcutaneous tissue as possible. If the specimen has been properly cleaned, the individual muscles should be visible. If, after thorough cleaning, the cat is covered with hairs, rinse it in *cold* water. Wrap up the skin and other discarded tissues in the plastic you have been working on, and put them in a waste can. Use new plastic and muslin for a working surface as well as for storage wraps for the cat.*

Note: It will be necessary, from time to time, to dip the cat in a preservative solution (usually 2–3% phenol) to keep it moist and well preserved. When storing the cat, be sure that it is securely wrapped, first in muslin and then in plastic.

As you proceed with the dissection of muscles, keep in mind that *dissect* means "to separate"—not to chop or dig!

A thin, tough sheet of fascia covers the muscles and binds them together. Where this fascia is so thick that it obscures the muscle fibers, it will be necessary to remove it, but *do not* remove the **lumbodorsal fascia** (Figs. 3-1, and 3-2, pp. 28 and 30).

Using a metal probe to break the connective tissue (use the scalpel only if necessary), separate the muscles at the cleavage lines, which should be visible if the cat has been properly cleaned. If these lines are not visible but fiber direction is apparent, pull the muscles apart slightly; this should cause the cleavage to appear. You should follow each muscle to origin and insertion as far as is practical. It will be necessary to transect superficial muscles in order to see deeper muscles, and instructions will be given when needed. In *transection* a cut is made transversely, usually at about the center of the belly of the muscle, so that enough fibers are left attached to both the origin and the insertion for later identification. Fascia under superficial muscles must be broken and removed in order to expose deeper muscles.

Dissect for deep muscles on one side only, and dissect the other side only superficially so that the superficial muscles are left intact.

In some courses, the human muscles are studied in considerable detail. For these classes a number of human skeletal diagrams, which are to be used for drawing the human muscles in place, are included at the end of this chapter. Where both right and left sides, or two or more outlines of the same side are included, use one for superficial muscles and the other(s) for deeper muscles. Indicate origin and insertion on the diagrams, and study these and muscle fiber direction to help you determine the kinds of movement produced by the muscles.

Muscle Groups of the Head, Neck, and Pectoral Appendage

Some muscles can be classified as belonging to more than one group. You will not observe some muscles of certain groups until you have studied other, more superficial muscles and transected them.

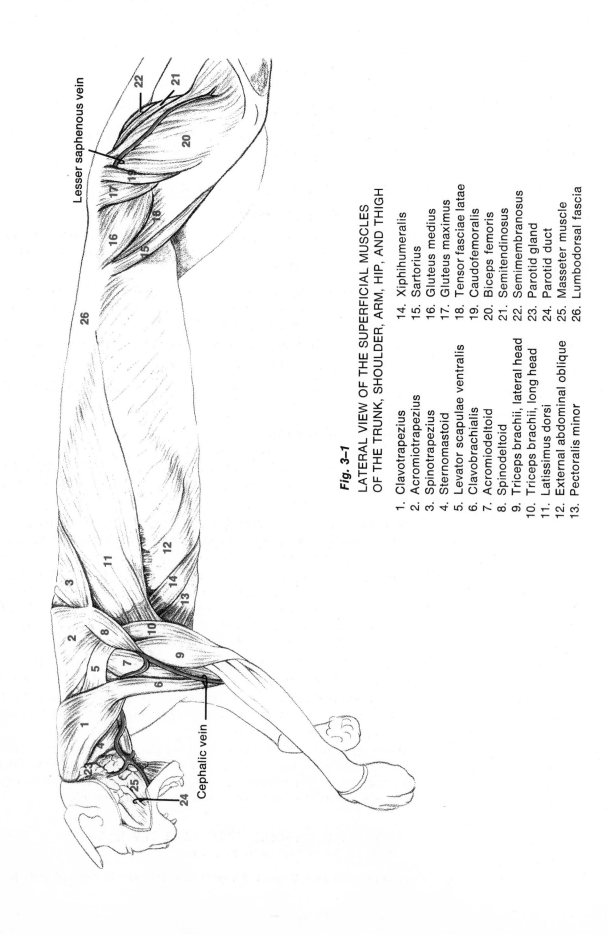

Lesser saphenous vein

Cephalic vein

Fig. 3–1

LATERAL VIEW OF THE SUPERFICIAL MUSCLES
OF THE TRUNK, SHOULDER, ARM, HIP, AND THIGH

1. Clavotrapezius
2. Acromiotrapezius
3. Spinotrapezius
4. Sternomastoid
5. Levator scapulae ventralis
6. Clavobrachialis
7. Acromiodeltoid
8. Spinodeltoid
9. Triceps brachii, lateral head
10. Triceps brachii, long head
11. Latissimus dorsi
12. External abdominal oblique
13. Pectoralis minor
14. Xiphihumeralis
15. Sartorius
16. Gluteus medius
17. Gluteus maximus
18. Tensor fasciae latae
19. Caudofemoralis
20. Biceps femoris
21. Semitendinosus
22. Semimembranosus
23. Parotid gland
24. Parotid duct
25. Masseter muscle
26. Lumbodorsal fascia

CUTANEOUS MUSCLES

You should have already observed the **cutaneous maximus** (a trunk muscle) and the **platysma** when you removed the skin.

MUSCLES OF THE HEAD AND FACE

Muscles of Facial Expression

This group is not usually studied in the cat. The muscles are essentially cutaneous muscles, and most are removed with the skin, but you may be able to observe the fibers of a few.

Occipitofrontalis, or **epicranius** (Fig. 3-2, p. 30). Note the broad, flat tendon over the cranium between the occipital and frontal portions of the muscle. This is the **galea aponeurotica.**

Orbicularis oculi. A sphincter muscle around the orbital opening.

Orbicularis oris. A sphincter muscle around the oral opening.

Note the cat's well developed muscles for moving the auricula, or pinna, of the ear.

Muscles of Mastication
(Figs. 3-1, 3-2, and 3-4, pp. 28, 30, and 35)

Note the location of the salivary glands (Figs. 3-3 and 3-4, pp. 32 and 35). Note the location and, insofar as is practical, the attachments of the muscles described below.

Masseter. A large rounded muscle in the cheek below the zygomatic arch, from which it arises. It inserts on the lateral side of the mandible. Note the duct of the parotid gland crossing the muscle.

Temporalis. A large fan-shaped muscle at the side of the skull, that arises from the parietal bone and squamous portion of the temporal bone. In order to see this muscle, slit the galea aponeurotica and reflect it. The temporalis passes medial to the zygomatic arch and inserts on the coronoid process of the mandible. Do not follow it to insertion.

Digastric(us). A long band-like muscle extending from the mastoid, or digastric groove to the ventral border of the mandible near the midline.

Mylohyoid(eus). Fibers course horizontally below the mandible, deep to the digastric. Origin is from the mandible, and insertion is into a median raphé and the hyoid bone.

Pterygoids, internal and **external.** (Pterygoideus externus, P. internus). The pterygoid muscles, which you will not observe unless the orbit is dissected, originate from the pterygoid processes of the sphenoid bone and insert on the inner surface of the body of the mandible.

SOME NECK MUSCLES THAT MOVE THE HEAD
(Figs. 3-1, 3-4, and 3-6, pp. 28, 35, and 38)

Sternocleidomastoid(eus). Extends from the sternum and clavicle to the

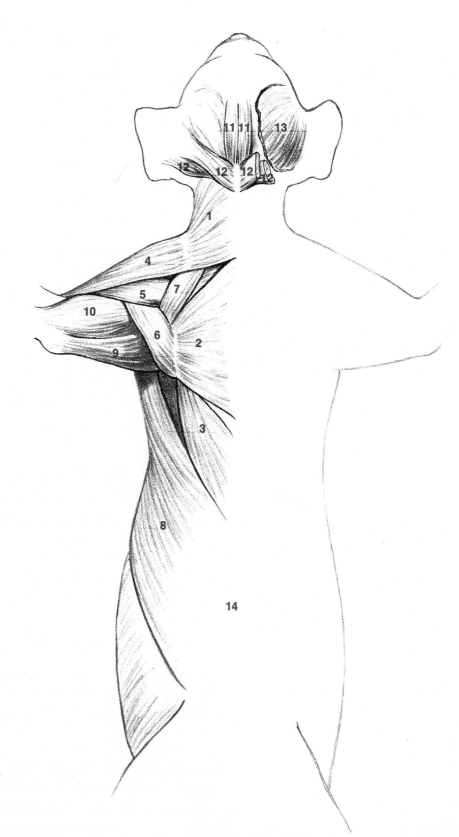

Fig. 3–2

SUPERFICIAL MUSCLES
OF THE BACK
(WHICH ATTACH THE
PECTORAL APPENDAGE
TO THE VERTEBRAL
COLUMN), AND SOME
MUSCLES OF THE
SHOULDER, ARM,
AND HEAD

1. Clavotrapezius
2. Acromiotrapezius
3. Spinotrapezius
4. Clavobrachialis
5. Acromiodeltoid
6. Spinodeltoid
7. Levator scapulae ventralis
8. Latissimus dorsi
9. Triceps brachii, long head
10. Triceps brachii, lateral head
11. Occipitofrontalis (occipital
    portion)
12. Muscles of auricula of the ear
13. Temporalis (ear muscles
    removed)
14. Lumbodorsal fascia

mastoid process. In the cat the sternocleidomastoid consists of two separate muscles:

Sternomastoid(eus). Note the ventrolateral position of this muscle in the neck. Remove enough fascia to see the muscle clearly, but do not damage the external jugular vein that crosses it and passes deep between the cleidomastoid and the sternomastoid. (In the human, the vein passes deep at the dorsal border of the sternocleidomastoid.)

Cleidomastoid(eus). Lateral to the sternomastoid, and deep to the ventral border of the clavotrapezius.

Splenius. You will not observe this muscle until you have studied and transected the trapezius. The splenius, which is deep to the trapezius, arises from the cervical ligament and inserts on the skull on a ridge between the occipital bone and the parietal bones. The human insertion is on the mastoid process and on the occipital bone just inferior to the superior nuchal line.

Trapezius. This muscle will be studied with another group.

Other muscles of this group will not be studied in the laboratory.

If your class is emphasizing the human muscles, draw in place, on Diagrams 1, 2a, and 3a, the human muscles corresponding to those you have just studied in the cat. On 2a and 3a, place the superficial muscles on one side and the deep muscles on the other.

SOME MUSCLES OF THE HYOID, LARYNX, AND TONGUE
(Figs. 3-3 and 3-4, pp. 32 and 35)

If the sternomastoid muscles are fused across the midline at their caudal ends, cut them apart and pull them aside in order to locate the muscles just ventral to the larynx and trachea. Muscle attachments can be determined from the names. The prefix *genio-* denotes the chin; the suffix *-glossus* denotes the tongue; *stylo-* refers to the styloid process of the temporal bone, *sterno-* to the sternum, *thyro-* and *-thyroid* to the thyroid cartilage of the larynx, *crico-* to the cricoid cartilage of the larynx, *hyo-* and *-hyoid* to the hyoid bone.

Do not disturb the bundle of blood vessels and nerves that lies laterad and dorsad of these muscles, on each side.

Sternohyoid(eus). A narrow, straight muscle at either side of the midline. Separate it from underlying muscles; transect and reflect on one side.

Sternothyroid(eus). Laterad and slightly dorsad of the sternohyoid. Take care not to damage the thyroid gland, which lies deep to this muscle.

Thyrohyoid(eus). Craniad of the sternothyroid and laterad of the sternohyoid.

Cricothyroid(eus). A small muscle deep to the sternohyoid.

Stylohyoid(eus). Note this tiny band crossing the outer surface of the digastric muscle to reach the hyoid bone.

Sever the transverse vein that crosses the neck ventrally, and transect the digastric on one side, taking care not to cut underlying structures. Pass a

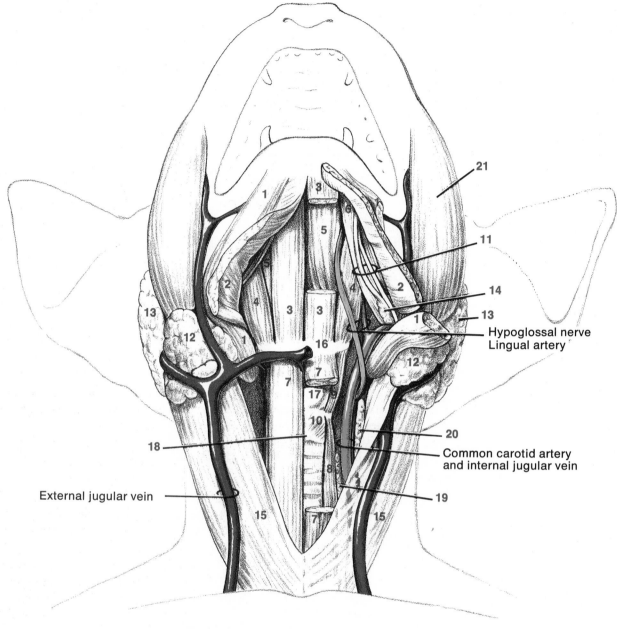

External jugular vein

Hypoglossal nerve
Lingual artery

Common carotid artery
and internal jugular vein

Fig. 3–3

MUSCLES OF THE HYOID AND LARYNX REGIONS,
AND EXTRINSIC MUSCLES OF THE TONGUE

1. Digastric
2. Mylohyoid (reflected)
3. Geniohyoid
4. Hyoglossus
5. Genioglossus
6. Styloglossus
7. Sternohyoid
8. Sternothyroid
9. Thyrohyoid
10. Cricothyroid
11. Ducts of submandibular
    and sublingual glands

12. Submandibular gland
13. Parotid gland
14. Sublingual gland
15. Sternomastoid muscle
16. Hyoid bone
17. Location of thyroid cartilage
18. Location of cricoid cartilage
19. Thyroid gland
20. Lymph node
21. Masseter

probe under the mylohyoid to loosen it; transect and reflect to find the following muscles:

Geniohyoid(eus). Located at either side of the midline. Separate it from underlying muscles; transect and reflect on one side.

Genioglossus. This lies deep to the cranial portion of the geniohyoid, and extends slightly laterad.

Hyoglossus. Lies laterad of the caudal part of the geniohyoid.

Styloglossus. Lies deep to the digastric, and craniad of the lingual artery and hypoglossal nerve.

MUSCLES THAT ATTACH THE PECTORAL APPENDAGE TO THE VERTEBRAL COLUMN
(Figs. 3-1, 3-2, and 3-6, pp. 28, 30, and 38)

Trapezius. A large thin muscle, triangular in shape, which arises from the occipital bone and thoracic spines, and inserts on the pectoral girdle. This muscle and the latissimus dorsi make up the superficial layer of muscle over the back region. In the cat, the trapezius has three parts:

> **Clavotrapezius.** This muscle is comparable to the clavicular portion of the human muscle. It attaches to the tiny clavicle and is continuous with the clavobrachialis, which extends into the arm region.

> **Acromiotrapezius.** This muscle is comparable to the part of the human muscle that inserts on the acromion. Note the flat, loose tendon that connects this muscle with its counterpart on the opposite side. This connection is not as loose in the human as in the cat.

> **Spinotrapezius.** This is comparable to the part of the human muscle that inserts on the scapular spine. It is superficial to the cranial border of the latissimus dorsi.

Latissimus dorsi. A flat, broad muscle, triangular in shape, with extensive origin in the back from spines of the last six thoracic vertebrae and all of the lumbar vertebrae, and from the sacrum and the iliac crest by aponeurosis (lumbodorsal fascia). Insertion is at the distal end of the intertubercular groove of the humerus.

Separate the trapezius and latissimus dorsi muscles from each other and from underlying structures by inserting the metal probe under them or by using your fingers. Transect and reflect the separate parts of the trapezius; also transect and reflect the latissimus dorsi.

Rhomboideus. A rhomboid-shaped muscle situated deep to the trapezius, passing from the spines of the upper thoracic vertebrae to the vertebral border of the scapula. The rhomboideus is one muscle in the cat. If two parts can be distinguished, the cranial part is larger and could be regarded as rhomboideus minor, while the small caudal part inserting at the inferior angle of the scapula might be regarded as rhomboideus major. The cranial-caudal position is the same as in the human, but r. major is larger than r. minor in the human.

Levator scapulae ventralis. Lies deep to the ventral portion of the clavotrapezius. It joins the ventral edge of the acromiotrapezius and inserts on the acromial part of the scapular spine. It arises from the transverse process of the atlas and the occipital bone. This muscle is not present in the human.

Occipitoscapularis (Rhomboideus capitis). A narrow band closely related to the rhomboideus. It arises from the occipital bone and inserts on the scapula just cranial to the insertion of the rhomboideus. It is superficial to the splenius, which can be observed at the side of the neck after superficial muscles have been transected. The occipitoscapularis is not present in the human.

To see the next muscle, pull the superior, or cranial border of the scapula from the body wall and rotate it forward; the cranial border of the muscle will appear between the inner surface of the scapula and the body wall.

Levator scapulae. A cranial continuation of the serratus anterior, with origin from transverse processes of the caudal five cervical vertebrae, and insertion on the vertebral border of the scapula, ventral to insertion of the rhomboideus. In the human, the insertion of the levator scapulae is comparable to that of the cat, but the origin (transverse processes of the first four cervical vertebrae) is roughly comparable to the origin of the cat's levator scapulae ventralis.

Draw the human muscles in place on Diagram 2a, putting the superficial muscles on one side and the deeper muscles on the other.

MUSCLES THAT ATTACH THE PECTORAL APPENDAGE TO THE VENTRAL AND LATERAL BODY WALL
(Figs. 3-4, 3-5, and 3-7, pp. 35, 36, and 40)

Pectoralis group (pectorales). Although the human has only two muscles in this group, the pectoralis major and minor, the cat has four:

Pectoantibrachialis. A narrow ribbon superficial to the pectoralis major, and closely applied to it. The cranial border parallels the ventral border of the clavobrachialis, which it merges with in the arm region.

Pectoralis major. Lies immediately deep to the pectoantibrachialis and, because it is larger, it can be seen below the caudal border of the other. The fibers of both muscles run in an essentially transverse direction from sternum to humerus. In the human the pectoralis major is larger, with a more extensive origin on the clavicle, the sternum, and the costal cartilages of the true ribs, but the insertion is less extensive than in the cat, being confined to a small area on the humerus just distal to the greater tuberosity.

Pectoralis minor. In the human this is a small muscle, but in the cat it is usually larger than the pectoralis major. In the cat it arises from the sternum, caudal to the origin of the pectoralis major, and inserts on both the humerus and the scapula. The fibers run diagonally craniad and laterad. In the human the pectoralis minor arises from ventral surfaces of the third, fourth, and fifth ribs, and inserts only on the coracoid process of the scapula.

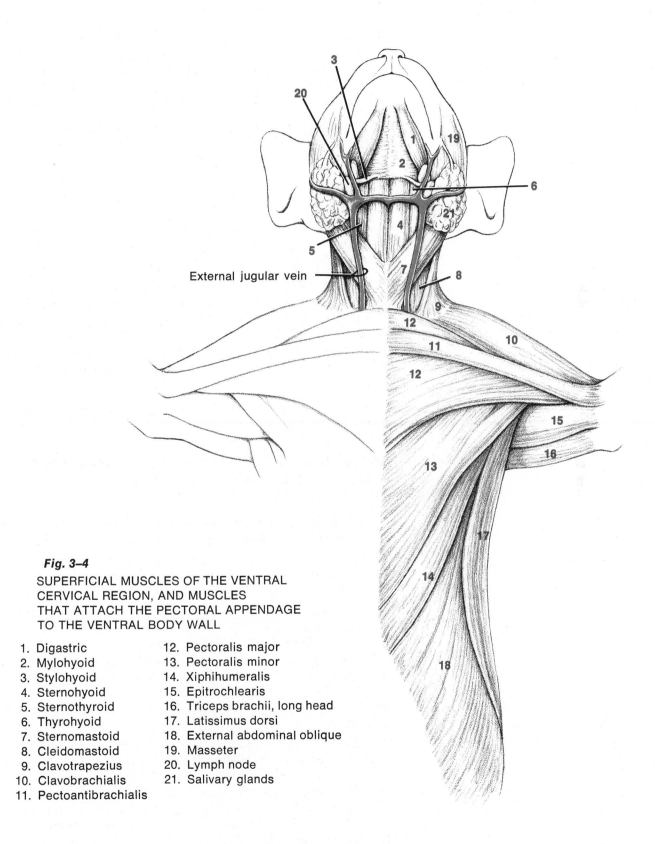

Fig. 3–4

SUPERFICIAL MUSCLES OF THE VENTRAL
CERVICAL REGION, AND MUSCLES
THAT ATTACH THE PECTORAL APPENDAGE
TO THE VENTRAL BODY WALL

External jugular vein

1. Digastric
2. Mylohyoid
3. Stylohyoid
4. Sternohyoid
5. Sternothyroid
6. Thyrohyoid
7. Sternomastoid
8. Cleidomastoid
9. Clavotrapezius
10. Clavobrachialis
11. Pectoantibrachialis

12. Pectoralis major
13. Pectoralis minor
14. Xiphihumeralis
15. Epitrochlearis
16. Triceps brachii, long head
17. Latissimus dorsi
18. External abdominal oblique
19. Masseter
20. Lymph node
21. Salivary glands

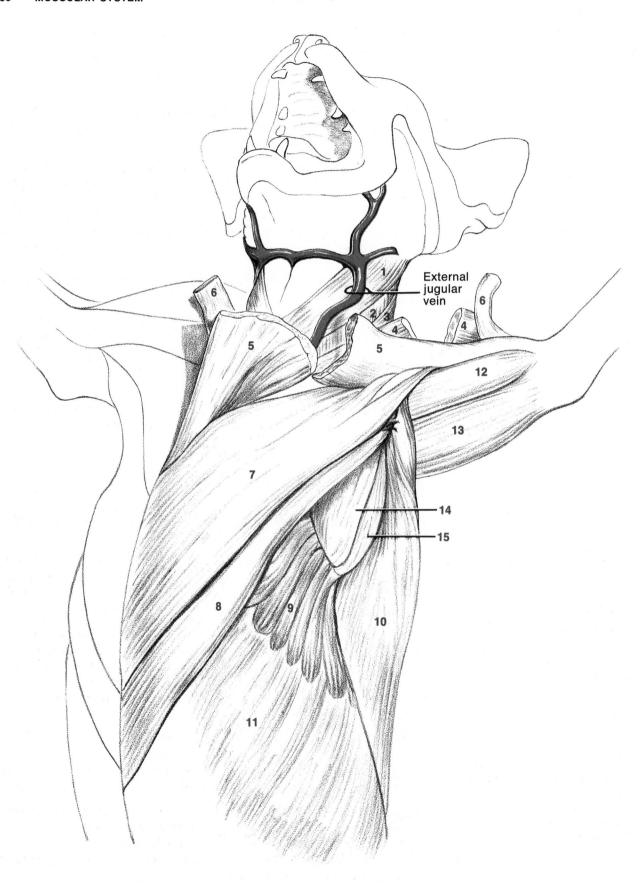

External
jugular
vein

Xiphihumeralis. The most caudal of the pectorales. It arises from the xiphoid region of the sternum and, along with the pectoralis minor and the latissimus dorsi, inserts on the humerus.

Separate each of the pectoralis muscles and transect each, being extremely careful not to destroy the brachial nerve plexus (Fig. 8-11, p. 142) and blood vessels that lie immediately deep to these muscles in the axilla. (The axilla is the space bounded ventrally by the pectorales, dorsally by the latissimus dorsi, teres major, and subscapularis, medially by the serratus anterior, and laterally by the arm muscles. "Armpit" is the name commonly applied to the axilla.)

Pull the scapula away from the body wall and rotate it dorsad. Clean off enough fascia so that the next muscle can be observed.

Serratus anterior, or **ventralis,** or **magnus.** Note the serrate margin due to the origin by digitations from the upper eight or nine ribs. The digitations merge into a compact muscle that runs laterad and craniad to the vertebral border of the scapula, to insert ventral to the rhomboideus. Note the cranial continuation as the levator scapulae.

Draw the human muscles in place on Diagram 3a, placing the superficial muscles on one side and deeper muscles on the other side.

MUSCLES OF THE SHOULDER
(Figs. 3-1, 3-2, 3-6, and 3-7, pp. 28, 30, 38, and 40)

Clean off as much fascia as necessary to expose muscle fibers, and then separate the muscles.

Deltoid(eus). In the human this is a thick, triangular muscle with three heads of origin: clavicular, acromial, and spinous (spine of scapula). Fibers from all heads converge to insert on the deltoid tuberosity of the humerus. In the cat there are three divisions of the deltoid:

Clavobrachialis. This is somewhat comparable to the clavicular portion of the human muscle, and is sometimes called the clavodeltoid(eus). It is continuous at its proximal end (the origin) with the clavotrapezius. Some of its fibers merge with the pectoantibrachialis, which it parallels, to insert in the superficial fascia of the forearm. Most of the insertion is with the brachialis on the ulna.

Acromiodeltoid(eus). Comparable to the acromial portion in the human.

Spinodeltoid(eus). Comparable to the spinous portion in the human. The acromiodeltoid and spinodeltoid insert together on the humerus.

Supraspinatus. Reflect the previously transected clavotrapezius and acromiotrapezius to their insertions in order to observe the supraspinatus. From its origin in the supraspinous fossa it passes ventral to the acromion, and craniad of the shoulder joint to reach the greater tuberosity of the humerus, where it inserts.

Infraspinatus. Transect the spinodeltoid and reflect. The infraspinatus can

Fig. 3–5
MUSCLES THAT ATTACH
THE PECTORAL APPENDAGE
TO THE VENTRAL AND
LATERAL BODY WALL

1. Sternomastoid
2. Cleidomastoid
3. Clavotrapezius
4. Clavobrachialis
5. Pectoralis major
6. Pectoantibrachialis
7. Pectoralis minor
8. Xiphihumeralis
9. Serratus anterior
10. Latissimus dorsi (turned back)
11. External abdominal oblique
12. Biceps brachii
13. Epitrochlearis
14. Subscapularis
15. Teres major

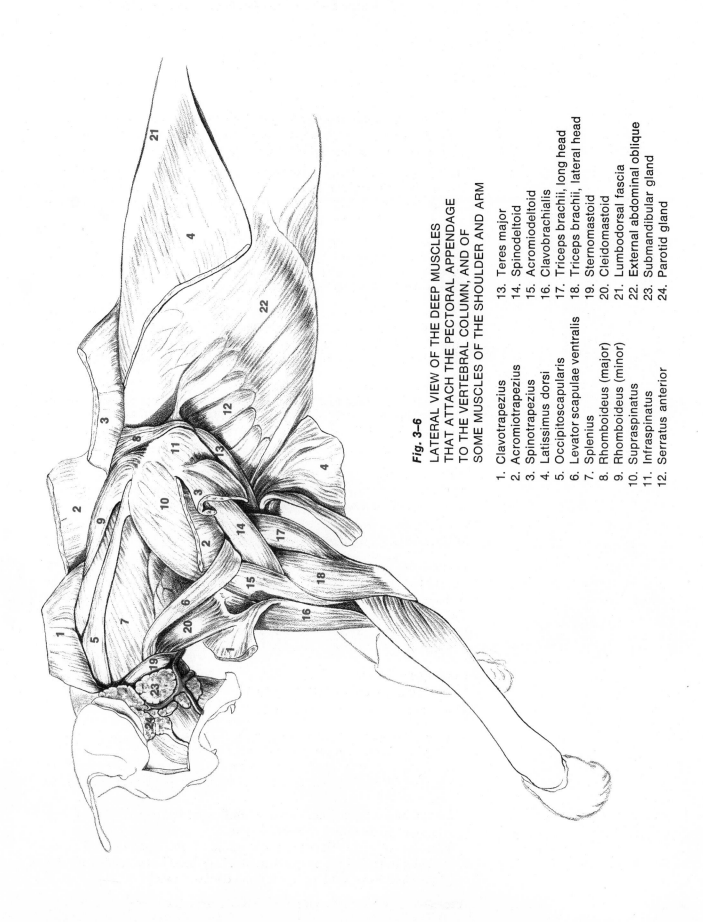

Fig. 3-6

LATERAL VIEW OF THE DEEP MUSCLES
THAT ATTACH THE PECTORAL APPENDAGE
TO THE VERTEBRAL COLUMN, AND OF
SOME MUSCLES OF THE SHOULDER AND ARM

1. Clavotrapezius
2. Acromiotrapezius
3. Spinotrapezius
4. Latissimus dorsi
5. Occipitoscapularis
6. Levator scapulae ventralis
7. Splenius
8. Rhomboideus (major)
9. Rhomboideus (minor)
10. Supraspinatus
11. Infraspinatus
12. Serratus anterior
13. Teres major
14. Spinodeltoid
15. Acromiodeltoid
16. Clavobrachialis
17. Triceps brachii, long head
18. Triceps brachii, lateral head
19. Sternomastoid
20. Cleidomastoid
21. Lumbodorsal fascia
22. External abdominal oblique
23. Submandibular gland
24. Parotid gland

now be observed extending from its origin in the infraspinous fossa, passing deep to the spinodeltoid, to its point of insertion on the greater tuberosity of the humerus, adjacent to insertion of the supraspinatus.

Teres major. From the axillary border of the scapula, caudal to the infraspinatus, this muscle passes ventral to the long head of the triceps brachii to insert with the latissimus dorsi on the humerus, along the medial margin of the intertubercular groove.

Teres minor. A small muscle from the axillary border of the scapula, positioned between the infraspinatus and the long head of the triceps brachii. Its insertion is just distal to that of the infraspinatus, on the greater tuberosity of the humerus.

Separate the pectorales and latissimus dorsi from the fibers of the cutaneous maximus at the axilla, and reflect to their insertions. Clean the fascia from the axilla, being very careful not to destroy the brachial nerve plexus and axillary blood vessels. The next muscle can now be observed.

Subscapularis. The muscle fibers arise from the surface of the subscapular fossa and converge to a tendon that passes deep to the tendon of origin of the biceps brachii, and to the coracobrachialis, to insert on the lesser tuberosity of the humerus. (The teres major can also be observed along the axillary border of the scapula.)

Draw the human muscles in place on Diagrams 1, 2a, and 3a. On diagrams 2a and 3a, place the superficial and deep muscles on opposite sides to correspond to the other muscle placements you have made.

MUSCLES OF THE ARM (UPPER ARM)

Remove the brachial fascia from the arm muscles and separate the muscles. Take care that the brachioradialis (Fig. 3-8, p. 42), which will be considered with muscles of the forearm, is not removed. This narrow muscle, which arises on the lateral side of the humerus and extends into the forearm, may appear to be nothing more than a strand of superficial fascia accompanied by a nerve (branch of the radial nerve).

Dorsal Group
(Figs. 3-6, 3-7, and 3-8, pp. 38, 40, and 42)

This is basically an extensor group.

Triceps brachii. This muscle has three heads of origin, which merge to insert on the olecranon process of the ulna:

Long head (middle head). Arises from the infraglenoid tubercle of the scapula. This is proportionately much larger in the cat than in the human.

Lateral head. Arises from the dorsal surface of the humerus, laterad and craniad of the radial, or spiral groove.

Medial head. Arises from the dorsal surface of the humerus, medial and caudal to the radial, or spiral groove. This head has a number of divisions in the cat, which will not be studied separately.

Anconeus. A very small, flat muscle arising from the dorsal surface of the

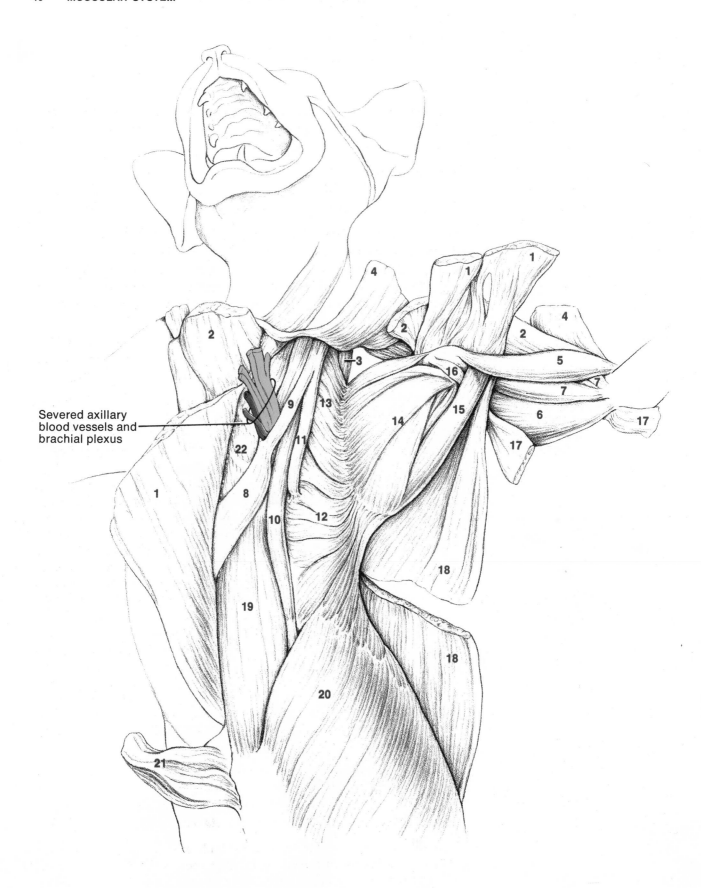

Severed axillary
blood vessels and
brachial plexus

lateral epicondyle of the humerus, and inserting on the olecrannon process of the ulna, immediately distal to the insertion of the triceps brachii.

Ventral Group

(Figs. 3-7 and 3-8, pp. 40 and 42)

This is basically a flexor and supinator group.

Clavobrachialis. A division of the deltoid muscle (see p. 37).

Pectoantibrachialis. One of the pectorales (see p. 34).

Epitrochlearis. This is not present in the human. It is a flat, thin, superficial muscle on the medial side of the arm. It arises from the latissimus dorsi and is continuous distally with the antibrachial fascia (fascia of the forearm).

Remove the epitrochlearis and the pectoantibrachialis to observe the remaining muscles, which are essentially the same as those found in the arm region of the human.

Coracobrachialis. In the cat this muscle is usually very small. Its origin, (coracoid process of the scapula) is the same as in the human, but its insertion on the humerus is usually more cranial in position in the cat. In the human, insertion is at about the middle of the medial surface of the humerus.

Biceps brachii. In the cat this muscle lies deep to the pectorales, just medial to their humeral insertions. The tendon of origin passes from a small tubercle above the glenoid fossa, through the capsule of the shoulder joint, and along the intertubercular groove. Insertion is on the bicipital tuberosity of the radius. The human biceps brachii has two heads of origin: a long head, which is comparable to the cat muscle, and a short head, which arises from the coracoid process of the scapula and joins the distal end of the long head. The insertion is the same in the human as in the cat.

Brachialis. In the cat this muscle arises from the lateral side of the humerus, crosses the cubital, or antecubital fossa (ventral depression at the elbow joint), and inserts on the ventral surface of the ulna just distal to the coronoid process. In the cat the insertion of the pectoralis major separates the biceps brachii and brachialis, the brachialis being on the lateral side. In the human the brachialis arises from the distal half of the ventral surface of the humerus, so that medially it lies deep to the biceps brachii.

Draw the human arm muscles in place on Diagrams 4 and 5, using *a* for superficial muscles and *b* for deeper muscles.

MUSCLES OF THE FOREARM AND HAND

Remove the remaining skin from the forelimb. Note and remove the tough sheath of fascia (antibrachial fascia) that encases the forearm muscles. Observe the long tendons of insertion of most of the muscles, and note that tendons extending into the hand are bound down at the wrist by transverse

Fig. 3–7
DEEP MUSCLES
THAT ATTACH THE
PECTORAL APPENDAGE
TO THE VENTRAL AND
LATERAL BODY WALL,
AND SOME SHOULDER
AND ARM MUSCLES

1. Pectoralis minor
2. Pectoralis major
3. Splenius
4. Clavobrachialis
5. Biceps brachii
6. Triceps brachii, long head
7. Triceps brachii, medial head
8. Transversus costarum
9. Scalenus anterior
10. Scalenus medius
11. Scalenus posterior
12. Serratus anterior
13. Levator scapulae
14. Subscapularis
15. Teres major
16. Coracobrachialis
17. Epitrochlearis
18. Latissimus dorsi
19. Rectus abdominis
20. External abdominal oblique
21. Xiphihumeralis
22. Internal intercostal

Fig. 3–8

SUPERFICIAL DORSAL
MUSCLES OF THE FOREARM,
AND SOME OF THE
SHOULDER AND ARM
MUSCLES (LEFT SIDE)

1. Latissimus dorsi
2. Triceps brachii, long head
3. Triceps brachii, medial head
4. Triceps brachii, lateral head
5. Brachioradialis
6. Brachialis
7. Spinodeltoid
8. Acromiodeltoid
9. Pectoralis major
10. Clavobrachialis
11. Extensor carpi radialis longus
12. Extensor carpi radialis brevis
13. Extensor digitorum communis
14. Extensor digitorum lateralis
15. Extensor carpi ulnaris
16. Anconeus
17. Antibrachial fascia
18. Extensor pollicis brevis

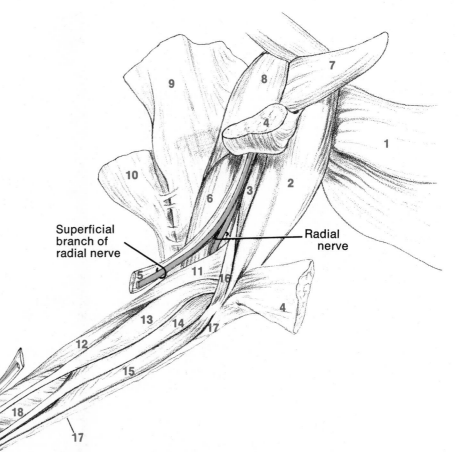

carpal ligaments (dorsal and ventral, or volar). It will be necessary to sever these ligaments to follow the tendons.

It is not necessary to transect all of the muscles of the forearm. If the forearm is dissected properly, you can observe most of the deep muscles by moving aside other muscles.

Dorsal Group: Superficial Muscles

(Fig. 3-8, p. 42)

The dorsal muscles are basically extensors and supinators. The superficial muscles are listed in order, from the radial to the ulnar side.

Brachioradialis. In the cat this is a narrow ribbon, which may have been removed with the superficial fascia. It arises with the dorsal group on the lateral side of the humerus and is supplied by the same nerve (radial), but it passes ventrad along the radial side of the forearm. It inserts on the lateral side of the styloid process of the radius. The brachioradialis is a supinator, but it is a flexor of the elbow joint rather than an extensor.

Extensor carpi radialis longus. Origin is from the lateral supracondylar ridge of the humerus, distal to the origin of the brachioradialis, and insertion is at the base of the second metacarpal.

Extensor carpi radialis brevis. Origin is from the lateral epicondyle of the

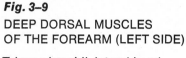

Fig. 3–9

DEEP DORSAL MUSCLES
OF THE FOREARM (LEFT SIDE)

1. Triceps brachii, lateral head
2. Triceps brachii, long head
3. Brachialis
4. Extensor carpi radialis
   longus
5. Extensor digitorum
   communis
6. Extensor digitorum lateralis
7. Supinator
8. Extensor pollicis brevis
9. Extensor indicis proprius
10. Extensor carpi ulnaris
11. Antibrachial fascia
12. Tendons of extensores carpi
    radialis longus and brevis

humerus by a tendon common to the origins of other extensor muscles, and insertion is at the base of the third metacarpal.

Extensor digitorum communis. Origin is from the lateral epicondyle by the common extensor tendon, and insertion is at the base of the middle and distal phalanges of all digits except the thumb.

Extensor digitorum lateralis. Origin is just distal to that of the extensor digitorum communis, to which it is supplementary. The tendons pass to the digits and join tendons of the extensor digitorum communis. The part of this muscle that gives a tendon to the fifth digit is comparable to the **extensor digiti quinti proprius** of the human, which is not present in the cat as a separate muscle. The extensor digitorum lateralis is not present in the human.

Extensor carpi ulnaris. Arises from the common extensor tendon and also from the dorsal border of the ulna. Insertion is on the medial side at the base of the fifth metacarpal.

Dorsal Group: Deep Muscles

(Fig. 3-9, p. 43)

Pull aside overlying muscles to locate deep muscles.

Extensor indicis proprius. Origin is from the middle of the ulna and the interosseous membrane; the tendon of insertion passes to the index finger and joins the tendon of the extensor digitorum communis. In the cat a tendon from the extensor indicis proprius goes to the thumb. This tendon corresponds to the **extensor pollicis longus** of the human, which is not present as a separate muscle in the cat.

Extensor pollicis brevis. The origin of this muscle is more extensive in the cat than in the human, since it comes from the ulna as well as from the dorsal surface of the radius and the interosseous membrane. The muscle corresponds to both the **extensor pollicis brevis** and the **abductor pollicis longus** of the human, the latter being absent as a separate muscle in the cat. In the cat, insertion is on the radial side of the base of the first metacarpal, which is the point of insertion of abductor pollicis longus in the human. In the human the extensor pollicis brevis inserts at the base of the proximal phalanx of the thumb.

Supinator. Origin is from the lateral epicondyle of the humerus, the radial collateral and annular ligaments, and the medial side of the ulna below the radial notch. Fibers pass obliquely distad to insert on the lateral and ventral sides of the proximal third of the radius.

It will be necessary to pull aside the superficial muscles near their origins in order to observe this muscle. (To observe the entire supinator it would be necessary to transect and reflect the overlying muscles. This is not desirable unless a good superficial dissection is done and left intact on the opposite forearm.)

Ventral Group: Superficial Muscles
(Fig. 3-10, p. 45)

The ventral muscles are basically flexors and pronators. The muscles are listed in order, from the radial to the ulnar side, for the superficial group.

Pronator teres. Origin is from the medial epicondyle of the humerus by a tendon that gives origin to other flexor muscles, and also from the medial side of the coronoid process of the ulna. It inserts on the middle third of the radius.

Flexor carpi radialis. Origin is from the medial epicondyle by common flexor tendon, and insertion is at the base of the second and third metacarpals.

Palmaris longus. Origin is from the medial epicondyle by common flexor tendon, and insertion is into the fascia of the palm. In the cat this muscle sends tendons to the digits, and it is larger than that in the human. (The muscle is not always present in the human. If it is present the tendon will stand out when the hand is flexed, since the tendon is not confined by the transverse carpal ligament.)

Flexor digitorum sublimis, or **superficialis.** In the cat this muscle consists of two parts, each part having a muscular origin. One part arises from the tendon of the palmaris longus, and one part from the ventral surface of the distal fleshy portion of the flexor digitorum profundus. The tendons pass to all digits except the thumb, and each tendon splits to insert at each side of the middle phalanx. In the human this is a larger muscle than in the cat and it lies deep to the palmaris longus and the flexores carpi. It has an extensive origin in the human: from the medial epi-

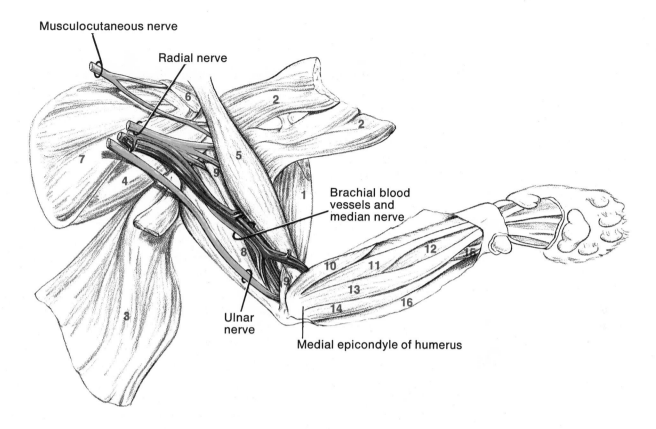

Musculocutaneous nerve

Radial nerve

Brachial blood vessels and median nerve

Ulnar nerve

Medial epicondyle of humerus

Fig. 3–10

SUPERFICIAL VENTRAL MUSCLES OF THE FOREARM, AND SOME OF THE ARM MUSCLES (LEFT SIDE)

1. Pectoralis major
2. Pectoralis minor
3. Latissimus dorsi and epitrochlearis (folded)
4. Teres major
5. Biceps brachii
6. Coracobrachialis
7. Subscapularis
8. Triceps brachii, long head
9. Triceps brachii, medial head
10. Pronator teres
11. Flexor carpi radialis
12. Flexor digitorum profundus
13. Palmaris longus
14. Flexor carpi ulnaris
15. Flexor digitorum sublimis, palmaris head
16 Antibrachial fascia

condyle by the common flexor tendon, from the medial side of the coronoid process of the ulna, and from the ventral surface of the radius distal to the insertion of the supinator.

Flexor carpi ulnaris. There are two heads of origin: a humeral head from the medial epicondyle by the common flexor tendon, and an ulnar head from the upper two-thirds of the medial border of the ulna. Insertion is on the pisiform and hamate carpal bones, and on the base of the fifth metacarpal.

Ventral Group: Deep Muscles
(Fig. 3-11, p. 46)

Transect and reflect the palmaris longus; pull aside other superficial muscles.

Flexor digitorum profundus. Origin is from the upper three-fourths of the ventral and medial surfaces of the ulna and from the interosseous membrane. In the cat this muscle has five heads of origin, which will not be studied separately. The muscle sends tendons, which are bound together as they cross the wrist, to the second, third, fourth, and fifth digits. Each tendon passes through the split tendon of the flexor digitorum sublimis and inserts at the base of the ventral surface of the distal phalanx. In the cat another tendon, which passes to the thumb, is given off, and this corresponds to the **flexor pollicis longus** of the human, which is not present as a separate muscle in the cat.

Pronator quadratus. A flat, quadrilateral muscle extending across the ventral surface of approximately the distal half of the radius and ulna. Origin is on the ulna, and insertion on the radius.

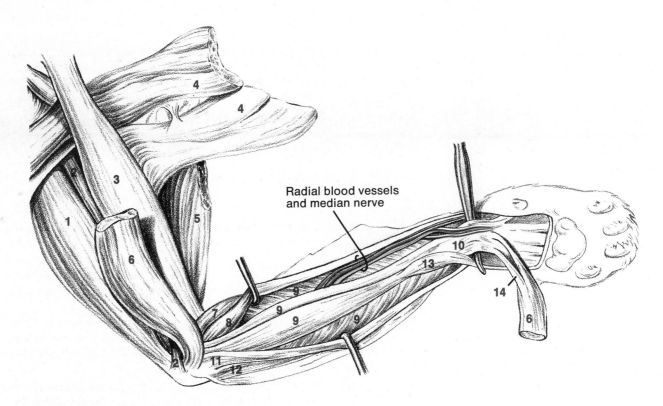

Radial blood vessels
and median nerve

Fig. 3–11
DEEP VENTRAL
MUSCLES OF
THE FOREARM
(LEFT SIDE)

1. Triceps brachii, long head
2. Triceps brachii, medial head
3. Biceps brachii
4. Pectoralis minor
5. Pectoralis major
6. Palmaris longus
7. Pronator teres
8. Flexor carpi radialis
9. Flexor digitorum profundus
10. Tendon bundle of flexor digitorum profundus
11. Flexor carpi ulnaris, humeral or long head
12. Flexor carpi ulnaris, ulnar or short head
13. Flexor digitorum sublimis, flexor profundus head
14. Flexor sublimis digitorum, palmaris head

In order to observe this muscle, which lies deep to the flexor digitorum profundus, you will have to move aside the tendon bundle of the latter, proximal to the wrist. (If it is desirable to view the entire muscle, it will be necessary to transect the flexor profundus tendon bundle proximal to the wrist and reflect the flexor profundus.)

A number of intrinsic hand muscles are present, but these will not be studied in the laboratory.

Draw the human muscles of the forearm in place on Diagrams 4 and 5, using *a* for superficial muscles and *b* for deeper muscles.

Other Features of the Pectoral Appendage

AXILLA, OR AXILLARY FOSSA

This space, which contains the brachial plexus of nerves and axillary blood vessels, is described on page 37.

CUBITAL, OR ANTECUBITAL FOSSA

This fossa is the depression ventral to the bend of the elbow. It is a triangular area, whose "floor" is formed by the brachialis and supinator. The lateral border is formed by the extensor carpi radialis muscles in the cat, but by the brachioradialis in the human. The pronator teres forms the medial border. In the cat the following structures pass into or through this triangle, from lateral to medial:

Radial nerve. Lies just medial to the extensor carpi radialis muscles. (In the human the radial nerve is just outside the fossa, under the brachioradialis.)

Tendon of insertion of the biceps brachii. Passes between the brachialis and the supinator at their insertions.

Brachial artery (and brachial vein in the human)

Median nerve

NERVES OF THE ARM, FOREARM, AND HAND
(Figs. 3-8, 3-10, and 3-11, pp. 42, 45, and 46)

Observe the major nerves. Nerves resemble white cords, the very small ones being difficult to distinguish from strands of connective tissue. In uninjected cats, nerves and arteries appear similar to one another. Identify the nerves by the muscles supplied.

Musculocutaneous. Passes to the coracobrachialis, biceps brachii, and brachialis. The name of the nerve indicates that it supplies both muscle and skin.

Radial. To the triceps brachii and anconeus, and all of the dorsal muscles of the forearm. You can observe this nerve where it crosses the ventral surface of the teres major and passes between parts of the medial head of the triceps brachii to the dorsal side of the arm.

You can also locate the radial nerve superficial to the lateral surface of the brachialis, by transecting the lateral head of the triceps brachii. Note the divisions into a superficial branch that passes into the superficial fascia and continues into the forearm and hand, and a deeper branch that extends to the muscles of the forearm, where it can be located passing between the supinator and the extensor carpi radialis muscles.

Ulnar. This nerve courses along the medial side of the long head of the triceps brachii and passes dorsal to the medial epicondyle and into the forearm region, where it can be found deep to the flexor carpi ulnaris, which it supplies. It also supplies the ulnar head of the flexor digitorum profundus and some intrinsic hand muscles.

Median. This nerve passes between the biceps brachii and the medial head of the triceps brachii on its course to the forearm. In the cat it passes through a foramen in the humerus, along with the brachial artery. (The brachial vein does not pass through this foramen.) In the forearm of the cat, the median nerve is accompanied by the radial blood vessels as it courses along the ventral surface of the flexor digitorum profundus. (In the human, the radial blood vessels are more laterad and are accompanied by a branch of the radial nerve.) The median nerve supplies all ventral muscles of the forearm that are not supplied by the ulnar nerve. It also supplies some of the intrinsic hand muscles.

BLOOD VESSELS OF THE ARM, FOREARM, AND HAND
(Figs. 3-10 and 3-11, pp. 45 and 46)

Observe the major blood vessels. Arteries are white in uninjected cats, red in injected cats. Veins, which have thinner walls than arteries, usually show brown spots in uninjected cats; they are blue in injected cats. It will be practical to find only the larger vessels in uninjected cats.

Arteries
Axillary. Can be located in the axilla.

Brachial. A continuation of the axillary artery into the arm region. Distal to the elbow this vessel becomes known as the radial artery. (In the human the brachial artery divides into radial and ulnar arteries just distal to the bend of the elbow.)

Radial. Passes along the ventral surface of the flexor digitorum profundus. It passes deep to the tendon of the extensor pollicis brevis and onto the dorsum of the hand. (In the human the radial blood vessels pass deep to the tendons of the abductor pollicis longus and extensores pollicis longus and brevis.)

Ulnar. A small branch of the radial artery. (In the human the ulnar artery is usually larger than the radial.) The ulnar and radial arteries anastomose in the hand region to form the palmar arch (two arches in the human: superficial and deep).

Veins

Cephalic. This superficial vein, which you observed when you removed the skin from the cat, courses craniad on the dorsal side of the pectoral extremity to the point at which it passes deep between the cleidomastoid and the levator scapulae ventralis, at the dorsal border of the clavotrapezius. It joins a deep vein, the transverse scapular. (In the human it joins the axillary vein.)

Median cubital. A branch of the cephalic vein that crosses the antecubital fossa to join the brachial vein. (In the human it joins the basilic vein, which is not present in the cat.)

Deep veins. These parallel the arteries and bear the same names: brachial, radial, and ulnar.

Muscle Groups of the Pelvic Appendage

Remove the fascia that encases the muscles of the hip and thigh so that muscle fiber direction can be determined. Take care not to remove the **iliotibial band**, or tract, which is a white tendinous band on the lateral side of the thigh. This band is connected with the deep fascia (fascia lata) of the thigh.

MUSCLES OF THE HIP REGION

Separate the muscles of the dorsal hip region (gluteal region) at the cleavage lines.

Dorsal Group: Superficial Muscles
(Figs. 3-1 and 3-12, pp. 28 and 49)

The dorsal hip muscles are basically extensors, abductors, and lateral rotators. (There are, of course, exceptions to this generalization.)

Tensor fasciae latae. On the lateral side of the hip region, ventral in position to other muscles of this group. It arises from the outer part of the iliac crest and anterior superior iliac spine, and from a notch below the spine. The fleshy part is comparatively short, with the ventral portion lying deep to the cranial extremity of the sartorius. The fibers terminate in the iliotibial band, which inserts on the lateral condyle of the tibia. This muscle is a flexor, abductor, and medial rotator.

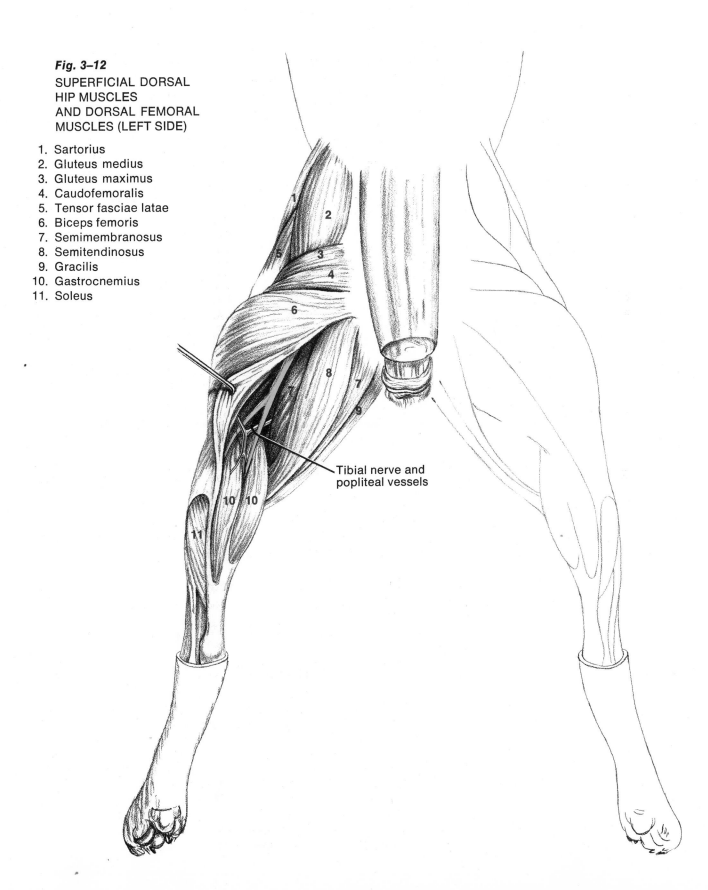

Fig. 3–12

SUPERFICIAL DORSAL
HIP MUSCLES
AND DORSAL FEMORAL
MUSCLES (LEFT SIDE)

1. Sartorius
2. Gluteus medius
3. Gluteus maximus
4. Caudofemoralis
5. Tensor fasciae latae
6. Biceps femoris
7. Semimembranosus
8. Semitendinosus
9. Gracilis
10. Gastrocnemius
11. Soleus

Tibial nerve and
popliteal vessels

Caudofemoralis. A small muscle arising on the caudal vertebrae and inserting by a very thin tendon on the patella. It is the most caudal one of the dorsal superficial hip muscles. It is not present in the human.

Separate the caudofemoralis from the gluteus maximus and sever the tendon of insertion of the former. Reflect the muscle to its origin.

Gluteus maximus. A very small muscle. It is craniad of Caudofemoralis, and its caudal border lies deep to it. Origin is from the last sacral and first caudal vertebrae and adjacent fascia; insertion is on the greater trochanter of the femur. The human muscle is much larger, with an extensive origin from the dorsal portion of the external surface of the ilium, the sacrum and coccyx, and fascia of spinal muscles; insertion is in the iliotibial band and upper dorsal surface of the femur. In the cat, the caudofemoralis and gluteus maximus together are more comparable to the human muscle than the gluteus maximus alone.

Sever the gluteus maximus at its insertion, and reflect it to its origin.

Gluteus medius. A thick muscle that arises from the external surface of the ilium between the anterior and inferior gluteal lines, and inserts on the greater trochanter. In the cat this muscle is larger than the gluteus maximus and will be found craniad of the latter.

Carefully separate the gluteus medius from the underlying muscles (the piriformis dorsally; the gluteus minimus ventrally) and from the tensor fasciae latae that is ventral to it. Transect and reflect.

Dorsal Group: Deep Muscles
(Figs. 3-13 and 3-14, pp. 51 and 52)

Piriformis. A small, flat, triangular muscle, from the ventral surface of the sacrum to the upper surface of the greater trochanter. In the human it is covered dorsally by the gluteus maximus, but in the cat by the gluteus medius. Note that the sciatic nerve passes deep to the piriformis and gluteus maximus and superficial to the gemelli muscles, the obturator internus, and the quadratus femoris.

Gluteus minimus. A small, somewhat pyramidal muscle lying deep to the ventral portion of the gluteus medius, with its dorsal border adjacent to the piriformis and the gemellus superior. In the human the muscle is more fan-shaped. The gluteus minimus arises from the ventral portion of the external surface of the ilium and inserts on the greater trochanter on its lateral side, slightly ventral and distal to the insertion of the gluteus medius.

On *one* specimen at each table, transect and reflect the piriformis in order to observe the next muscle.

Gemellus superior. A flat, triangular muscle arising, in the cat, from the dorsal border of the ischium and the ilium. Insertion is at the cranial extremity of the greater trochanter, near the insertion of the gemellus inferior and the obturator muscles. In the cat it lies deep to the piriformis, between the gluteus minimus and gemellus inferior. In the human it is caudal to the piriformis and cranial to the obturator internus, and the

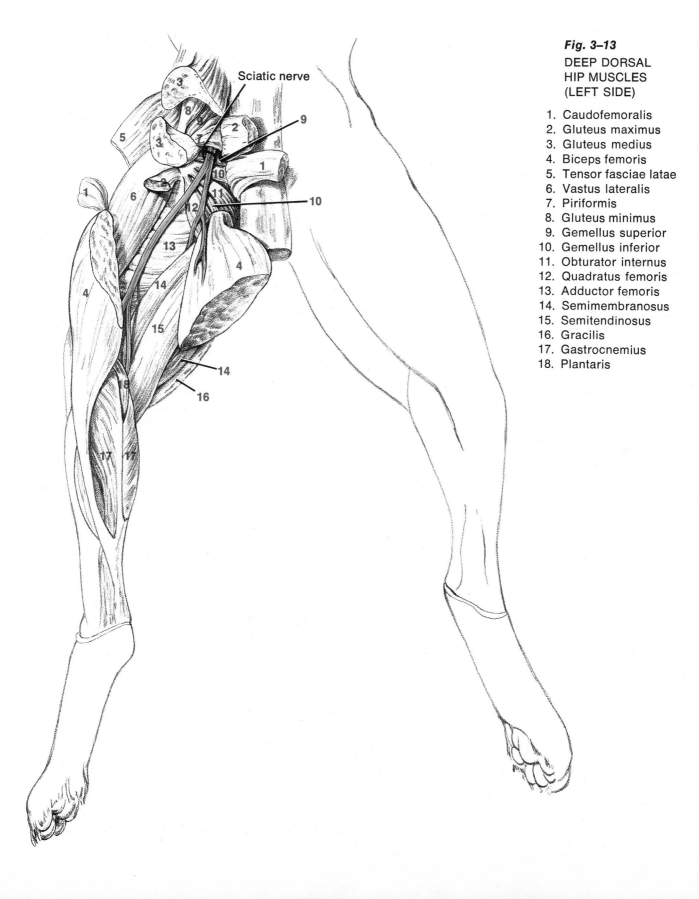

Sciatic nerve

Fig. 3–13
DEEP DORSAL
HIP MUSCLES
(LEFT SIDE)

1. Caudofemoralis
2. Gluteus maximus
3. Gluteus medius
4. Biceps femoris
5. Tensor fasciae latae
6. Vastus lateralis
7. Piriformis
8. Gluteus minimus
9. Gemellus superior
10. Gemellus inferior
11. Obturator internus
12. Quadratus femoris
13. Adductor femoris
14. Semimembranosus
15. Semitendinosus
16. Gracilis
17. Gastrocnemius
18. Plantaris

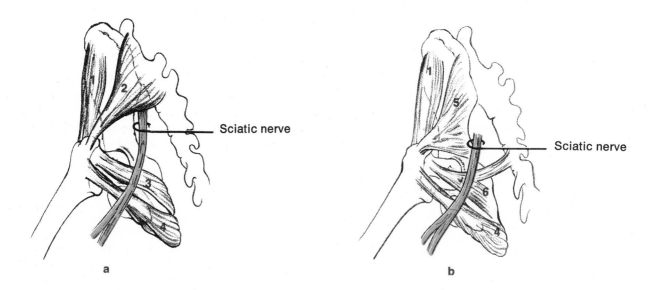

Sciatic nerve

Sciatic nerve

a

b

origin is less extensive, being only from the ischial spine.

Gemellus inferior. A flat, triangular muscle arising, in the cat, from the dorsal border of the ischium and the second or third caudal vertebrae. It is situated just caudal to the gemellus superior. Because most of it lies deep to the obturator internus, it cannot be completely observed unless the other is transected and reflected.

Maintain the obturator internus intact. The gemellus inferior inserts on the tendon of the obturator internus and, with the obturator internus, into the trochanteric fossa. In the human it lies caudal to the obturator internus, arising from only the ischial tuberosity.

Obturator internus. This muscle, lying caudal to the gemellus superior, arises from the inner surface of the ischium (from the tuberosity and the rami). It passes dorsally and "curves" over the lesser sciatic notch to reach the trochanteric fossa, where it inserts by a flat tendon.

Separate the fascial connections medial to the obturator internus to expose the "curving" fibers. Lift the tendon slightly with the probe and note the attachment of fibers of the gemellus inferior on the tendon.

Quadratus femoris. A short, thick muscle arising from the ischial tuberosity, caudal to the gemellus inferior. It inserts at the distal extremity of the greater trochanter, distal to insertions of the obturator and gemelli muscles. You can locate the quadratus femoris by lifting the ventral border of the proximal portion of the biceps femoris.

Obturator externus. A flat, triangular muscle, which you may not be able to observe. Near its origin, which is from the external surfaces of the inferior rami of the pubis and ischium and from the adjacent membrane over the obturator foramen, it lies deep to the adductor femoris. It passes dorsal to the hip joint to insert with the obturator internus and gemellus inferior. In some specimens it can be seen near its insertion by separating the adjacent borders of the gemellus inferior and the quadratus femoris. This muscle is an adductor rather than an abductor; otherwise its actions are similar to those of the other muscles of this group.

c

Fig. 3–14
ILLUSTRATIONS
SHOWING BOTH THE
CRANIAL-CAUDAL AND
THE DORSAL-VENTRAL
ALIGNMENT OF
THE DEEP DORSAL
HIP MUSCLES
(LEFT SIDE):
(a) shows the muscles
immediately underlying
the gluteus medius, gluteus
maximus, caudofemoralis,
and biceps femoris; in (b) the
piriformis and obturator
internus have been removed;
in (c) the quadratus femoris
has been removed to show
the ventral origin of the
obturator externus.

1. Gluteus minimus
2. Piriformis
3. Obturator internus
4. Quadratus femoris
5. Gemellus superior
6. Gemellus inferior
7. Obturator externus

Ventral Group
(Figs. 3-15 and 3-21, pp. 54 and 66)

The muscles of the ventral hip region (iliac region) have some origin craniad of this region, and help to form the posterior, or dorsal, abdominal wall; therefore, they are also considered to be posterior abdominal muscles. Description is provided here, but actual observation will be more practical after other abdominal muscles have been studied.

Iliopsoas. This muscle has two parts:
    Iliacus. Arises from the surface of the iliac fossa.
    Psoas major. Arises from the lumbar vertebrae. In the cat some fibers also arise from the tendons of the psoas minor.

Insertion of the two parts of the iliopsoas is on the lesser trochanter of the femur. The caudal extremity of the muscle can be observed at this time, medial to the sartorius and the rectus femoris.

Psoas minor. A small muscle arising from the bodies of the last thoracic vertebrae and the first four or five lumbar vertebrae. It inserts by a thin tendon onto an eminence at the junction of the ilium and superior ramus of the pubis. In the human this muscle arises from the last thoracic vertebra and the first lumbar vertebra, but it is absent in many humans.

Draw the hip muscles of the human in place on Diagrams 6 and 7, placing the superficial muscles on one side and the deep muscles on the other side.

MUSCLES OF THE THIGH (FEMORAL REGION)

Remove the remaining skin from the hindlimb, and clear away the loose fascia. Before beginning the dissection of the muscles of the thigh, observe the following superficial blood vessels:

    Greater saphenous vein. Courses craniad on the medial side of the leg and thigh, along with the saphenous artery and nerve, and joins the femoral vein.
    Lesser saphenous vein. Courses craniad on the dorsal side of the leg and thigh and passes deep between the caudofemoralis and the biceps femoris. It joins a tributary to the internal iliac vein.

Clear away as much deep fascia as necessary to expose the muscles without destroying the tendons of insertion. In clearing fat from the popliteal fossa (dorsal* to the knee), take care that nerves and blood vessels within the fossa are not destroyed.

Dorsal Femoral Group
(Figs. 3-12 and 3-13, pp. 49 and 51)

These muscles, known as the "hamstring" muscles, are extensors of the hip joint and flexors of the knee joint. They all arise from the ischial tuberosity.

---

*NOTE: Remember that the terms "dorsal" and "ventral" are used, in this manual, to correspond to those positions in the human.

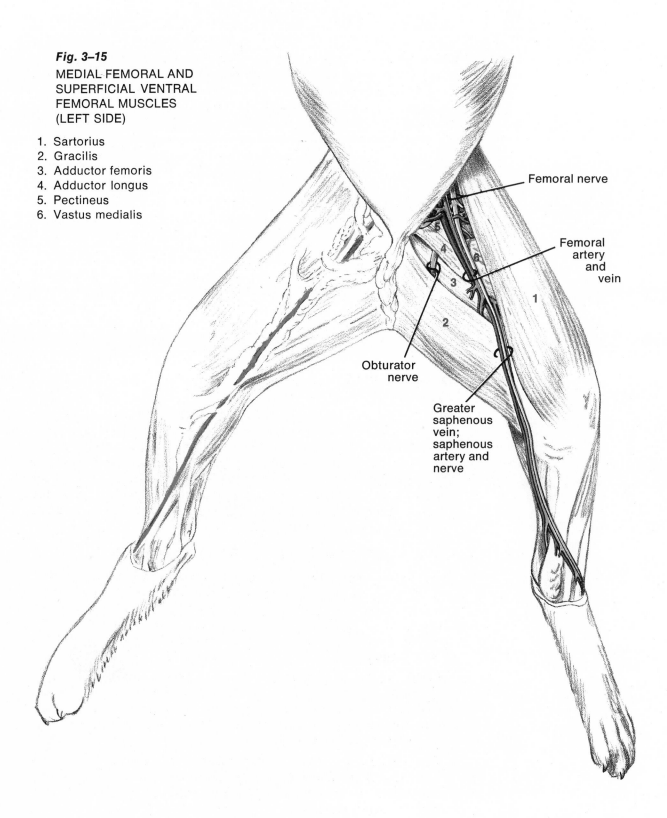

Fig. 3–15

MEDIAL FEMORAL AND
SUPERFICIAL VENTRAL
FEMORAL MUSCLES
(LEFT SIDE)

1. Sartorius
2. Gracilis
3. Adductor femoris
4. Adductor longus
5. Pectineus
6. Vastus medialis

Femoral nerve

Femoral
artery
and
vein

Obturator
nerve

Greater
saphenous
vein;
saphenous
artery and
nerve

Biceps femoris. A large muscle and the most lateral one of the group. In the cat, insertion is into the dorsolateral border of the proximal third of the tibia, and into the patella. In the human, insertion is onto the head of the fibula and onto the lateral side of the lateral condyle of the tibia.

A thin ribbon of muscle (the tenuissimus) arises near the caudofemoralis and passes along the inner surface of the biceps femoris. This muscle is not present in the human and can be disregarded.

Semimembranosus. The most medial one of the group. This is a large, thick muscle which inserts on the dorsal surface of the medial condyle of the tibia in the human, but in the cat it inserts both on the femur (medial epicondyle) and on the medial surface of the tibia.

Semitendinosus. The most superficial muscle of the group dorsally. It inserts into the dorsomedial border of the tibia, near its proximal end. Note the proximity of the insertions of the gracilis and sartorius muscles. Insertion is comparable in the human but extends farther ventrad, almost to the tibial tuberosity.

Ventral Femoral Group
(Figs. 3-15 and 3-16, pp. 54 and 56)

Sartorius. A strap-like muscle arising from the anterior superior iliac spine (origin is slightly more extensive than this in the cat). It passes diagonally across the ventral surface of the thigh, superficial to the quadriceps femoris, and inserts on the medial side of the tibia. In the human it passes just dorsal to the medial condyle of the femur; in the cat it passes medially and ventrally. Owing to this difference, it flexes the knee joint in the human, and extends it in the cat. In the cat there is also some insertion onto the patella.

Transect and reflect the sartorius and the iliotibial band.

Quadriceps femoris. This great extensor muscle of the knee joint is actually a group of muscles with a common insertion by a large tendon that extends to the patella, attaches around the patella, and passes on to the tibial tuberosity to insert. The portion of the tendon between the patella and tibia is called the **patellar ligament.** There are four divisions of this muscle:

Rectus femoris. A cigar-shaped muscle that arises from the anterior inferior iliac spine and also from an area just craniad of the acetabulum. It is bordered laterally by the vastus lateralis and medially by the vastus medialis. The rectus femoris is a flexor of the thigh as well as an extensor of the leg.

Vastus lateralis. A large muscle arising from the lateral and dorsal surfaces of the femur (along the lateral edge of the linea aspera) and from the greater trochanter. It is ventral to the biceps femoris, and in part is also medial.

Vastus medialis. A large muscle arising from the dorsal surface of the femur (along the medial edge of the linea aspera). It lies medial

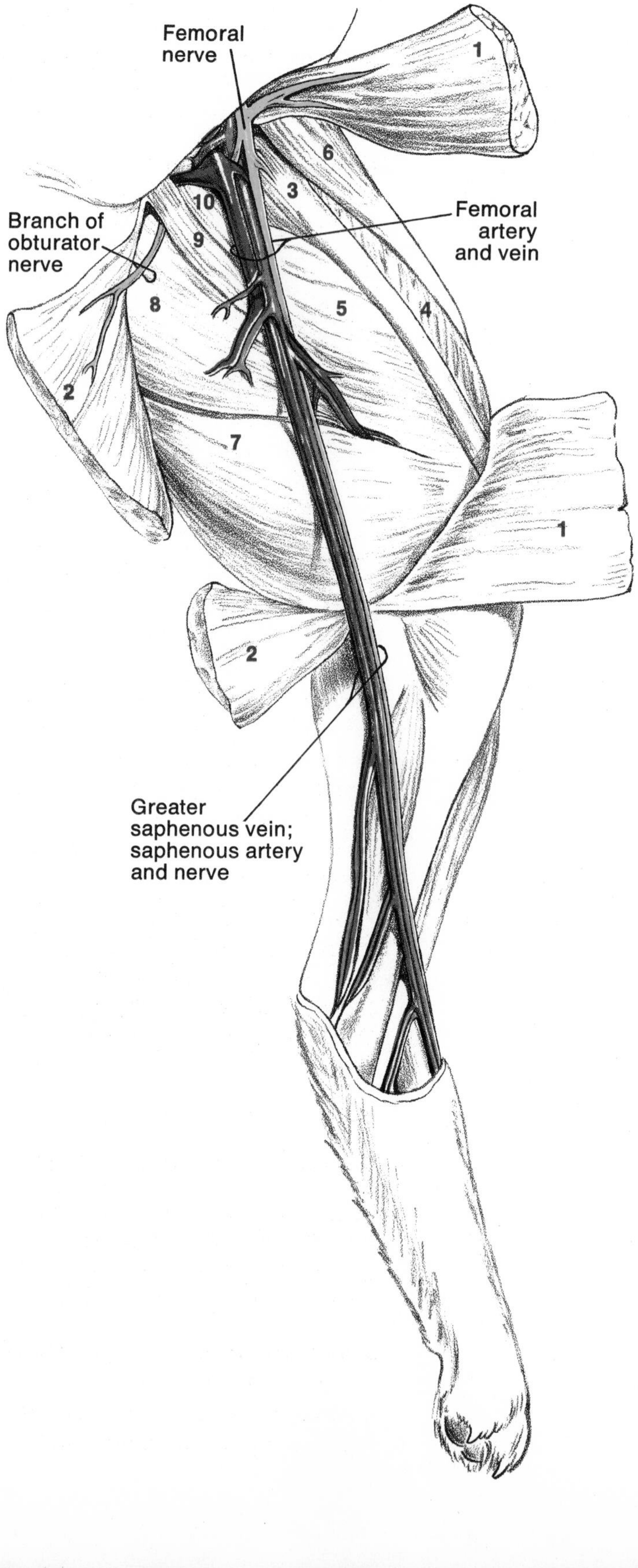

Femoral nerve

Branch of obturator nerve

Femoral artery and vein

Greater saphenous vein; saphenous artery and nerve

Fig. 3–16

MEDIAL FEMORAL
AND DEEP VENTRAL
FEMORAL MUSCLES
(LEFT SIDE)

1. Sartorius
2. Gracilis
3. Rectus femoris
4. Vastus lateralis
5. Vastus medialis
6. Tensor fasciae latae
7. Semimembranosus
8. Adductor femoris
9. Adductor longus
10. Pectineus

to the rectus femoris and lateral to the medial femoral muscles.

Vastus intermedius. A flat muscle deep to the rectus femoris, and between the vastus medialis and the vastus lateralis. The origin is from the ventral surface of the femur.

Medial Femoral Group
(Figs. 3-15 and 3-16, pp. 54 and 56)

This is the adductor group of the thigh.

Gracilis. A flat, wide band situated on the medial side of the thigh. Origin is from the inferior ramus of the pubis; in the cat the origin also extends to the ischium. In the cat the fibers end in a very thin, flat tendon, part of which becomes continuous with the fascia covering the distal portion of the leg. In the human, insertion is near the tibial tuberosity, with the sartorius and semitendinosus.

Pectineus. A small, flat muscle medial to the iliopsoas and vastus medialis, and lateral to the adductor longus. Origin is on the pubis, and insertion is on the dorsal surface of the femur just distal to the lesser trochanter (distal to the insertion of the iliopsoas).

Adductor longus. A thin muscle (but thicker than the pectineus) lying between the pectineus and the adductor femoris. Origin is from the cranial border of the pubis, and insertion is into the linea aspera at about the middle third. In the human, the adductor longus is superficial to the **adductor brevis** and to part of the **adductor magnus**, which are not present as such in the cat.

Adductor femoris. A large muscle lying between the adductor longus and the semimembranosus. Origin is from the inferior rami of both the pubis and the ischium; insertion is into the shaft of the femur throughout its length, along the linea aspera. This muscle corresponds to the adductores magnus and brevis of the human.

Draw the human muscles of the thigh in place on Diagrams 6, 7, and 8. On Diagrams 6 and 7 place the superficial muscles on one side and the deep muscles on the other. On Diagram 8 draw superficial muscles on *a*, and deep muscles on *b*. You can also place some of the hip muscles on Diagram 8, drawing the superficial muscles on *a*, and the deep muscles on *b*.

MUSCLES OF THE LEG AND FOOT

Remove the remaining skin from the foot and clean the superficial fascia from the leg and foot. Separate the muscles of the leg (crural region) by groups. It will be necessary to break some of the attachments of the femoral muscles to the fascia of the leg.

Dorsal Crural Group: Superficial Muscles
(Fig. 3-17, p. 58)

Gastrocnemius. Has two heads of origin: a lateral head from the lateral epicondyle of the femur, and a medial head from the medial epicon-

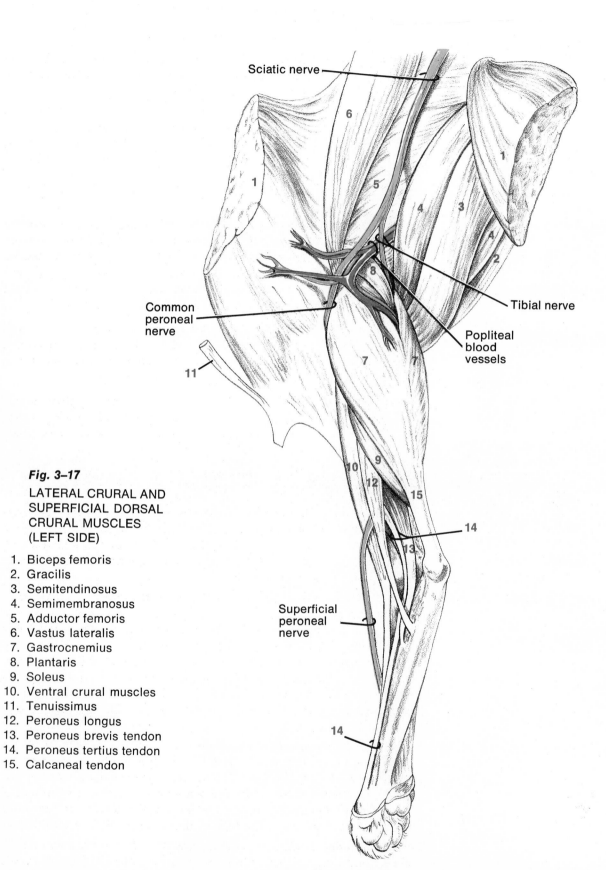

Sciatic nerve

Common peroneal nerve

Tibial nerve

Popliteal blood vessels

Superficial peroneal nerve

Fig. 3–17

LATERAL CRURAL AND
SUPERFICIAL DORSAL
CRURAL MUSCLES
(LEFT SIDE)

1. Biceps femoris
2. Gracilis
3. Semitendinosus
4. Semimembranosus
5. Adductor femoris
6. Vastus lateralis
7. Gastrocnemius
8. Plantaris
9. Soleus
10. Ventral crural muscles
11. Tenuissimus
12. Peroneus longus
13. Peroneus brevis tendon
14. Peroneus tertius tendon
15. Calcaneal tendon

of the interosseous membrane and the adjacent surfaces of the tibia and fibula. Muscle fibers converge to a tendon that passes deep to the flexor digitorum longus tendon and emerges ventral to it. Insertion is on the plantar surface of the foot.

The above three muscles are extensors of the foot. The flexor digitorum longus and flexor hallucis longus flex the digits, and the tibialis posterior inverts the foot.

Popliteus. A triangular muscle deep to the proximal portion of the gastrocnemius. It arises just distal to the origin of the lateral head of the gastrocnemius and fans out to insert on the medial side of the dorsal surface of the tibia, proximal to the popliteal line. The popliteus forms the "floor" of the popliteal fossa (see page 63) and is crossed by the tibial nerve and the popliteal blood vessels.

Lateral Crural Group
(Fig. 3-17, p. 58)

Peroneus longus. A slender, fusiform muscle situated at the lateral side of the leg, where it arises from the proximal half of the fibula. In the human, origin is also partly from the tibia. In the cat the tendon passes across the ventrolateral surface of the lateral malleolus and crosses the tendons of the peroneus tertius and peroneus brevis, to reach the plantar surface of the foot, where it inserts at the base of the metatarsals. In the human the tendon does not insert on all metatarsals, but only on the first, and on the first or second cuneiform.

Peroneus brevis. This muscle arises from the distal half of the fibula in the cat, but from the middle third in the human. It ends in a thick tendon which, in the cat, passes across the dorsolateral surface of the lateral malleolus in a groove, along with the tendon of the peroneus tertius. Insertion, in both cat and human, is on the lateral side of the tuberosity of the fifth metatarsal.

Peroneus tertius. A slender, fusiform muscle that arises from about the middle of the lateral surface of the fibula. In the cat, its tendon crosses the dorsolateral side of the lateral malleolus in a groove, along with the tendon of the peroneus brevis, and inserts on the tendon of the extensor digitorum longus that extends to the fifth digit. In the human, the peroneus tertius is a part of the ventral crural group, arising from the distal third of the ventromedial surface of the fibula. It is closely adjacent to the extensor digitorum longus and is often considered to be a part of it. The human tendon inserts at the superior surface of the base of the fifth metatarsal.

The peroneus muscles are everters and abductors of the foot, and the peroneus tertius extends and abducts the small toe. These muscles also assist with flexion and extension of the foot, but there is some variation between the cat and the human, owing to a difference in location of the tendons as they cross the ankle joint. In the human, the tendons of the peroneus longus and brevis pass behind the lateral malleolus and both muscles help to extend the foot. In the cat, the tendon of the peroneus

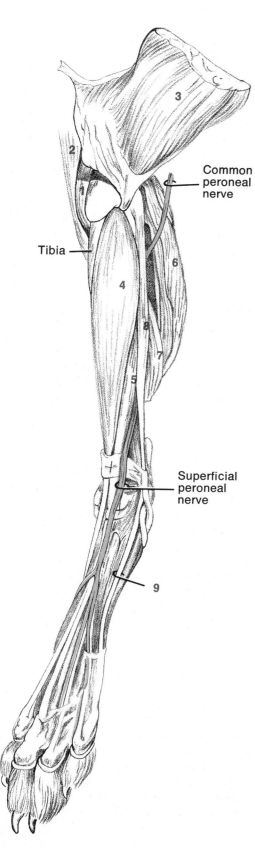

Tibia

Common peroneal nerve

Superficial peroneal nerve

Fig. 3–19
VENTRAL CRURAL
MUSCLES (LEFT SIDE)

1. Sartorius
2. Gracilis
3. Biceps femoris
4. Tibialis anterior
5. Extensor digitorum longus
6. Gastrocnemius, lateral head
7. Soleus
8. Peroneus longus
9. Tendon of peroneus tertius

longus passes across the ventrolateral surface of the lateral malleolus, and the muscle helps to flex the foot; the tendon of the peroneus brevis passes dorsolateral and the muscle helps to extend the foot. In the human, the peroneus tertius, with its tendon crossing the ankle joint ventrally, assists with flexion of the foot. In the cat, the peroneus tertius tendon crosses the ankle joint in a groove on the lateral malleolus, along with the tendon of the peroneus brevis; the muscle is small and any action produced, aside from that on the small toe, is minimal.

Ventral Crural Group
(Fig. 3-19, p. 62)

The muscles of this group are the flexors of the ankle joint. The tibialis anterior also inverts the foot, and those that send tendons to the digits extend the digits.

Tibialis anterior. The most superficial muscle of this group. Origin is on about the proximal half of the lateral surface of the tibia and adjacent interosseous membrane. The tendon crosses the ankle obliquely to reach the medial surface of the first metatarsal, where it inserts. In the human it inserts also on the first cuneiform.

Extensor digitorum longus. The origin of this muscle differs between cat and human. In the cat, origin is by a flat tendon from the lateral epicondyle of the femur. In the human, origin is from the lateral surface of the lateral tibial condyle and the ventromedial surface of the upper two-thirds of the fibula and the interosseous membrane. The tendon crosses the ankle ventrally and divides into four tendons that are distributed to all digits in the cat, and to all but the big toe in the human. Insertion is at the base of each distal phalanx.

In the human, as mentioned above, the peroneus tertius belongs to the ventral group. The human has another ventral crural muscle that is not present in the cat. This is the **extensor hallucis longus**, which arises on the middle half of the fibula and adjacent interosseous membrane, and passes to the base of the distal phalanx of the big toe to insert. The muscle lies between the tibialis anterior and the extensor digitorum longus.

There are a number of intrinsic muscles of the foot that will not be studied in the laboratory.

Draw the human crural muscles in place on Diagrams 8, 9, and 10. Place the lateral crural muscles on Diagram 8. The dorsal muscles should be placed on Diagram 9: the superficial muscles on *a* and *b*, the deep muscles on *c*. On Diagram 10, place the most superficial ventral muscles on *a* and the deeper ventral muscles on *b*. Place as many lateral crural muscles on both 9 and 10 as you can without interfering with placement of the other muscles.

Other Features of the Pelvic Appendage

FEMORAL TRIANGLE
(Fig. 3-15, p. 54)
Note the triangle formed by the medial border of the sartorius and the lateral border of the gracilis, with the other medial femoral muscles providing the "floor." Find, from lateral to medial, the following: femoral nerve, femoral artery, femoral vein.

POPLITEAL FOSSA
(Figs. 3-17 and 3-18, pp. 58 and 60)
This is the space dorsal to the knee joint at the distal end of the femur. It is bounded by the biceps femoris laterally, the semimembranosus medially, and the plantaris and the heads of the gastrocnemius distally. The "floor" is formed by the popliteus. This space contains the popliteal artery and vein, the tibial nerve, lymph nodes, connective tissue, and fat.

NERVES OF THE THIGH, LEG, AND FOOT
(Figs. 3-15, 3-16, 3-17, and 3-18, pp. 54, 56, 58, and 60)

Femoral. Emerges from the psoas major and distributes to the ventral femoral muscles. It gives off a cutaneous branch, the saphenous nerve, which passes to the leg and foot.

Obturator. Emerges from the obturator foramen and supplies the medial femoral muscles.

Sciatic. Follow this nerve from the hip, where it crosses the quadratus femoris (deep to the biceps femoris), to its division into the tibial and common peroneal nerves. Note the branches to the dorsal femoral muscles, all of which the sciatic nerve supplies.

Tibial. Passes through the popliteal fossa, courses deep to the superficial dorsal crural muscles, and passes into the foot. This nerve supplies all of the dorsal crural muscles and some of the intrinsic foot muscles.

Common peroneal. Divides into a deep peroneal branch, which supplies the ventral crural muscles, and a superficial peroneal branch, which supplies the lateral crural muscles. These branches continue into the foot to supply intrinsic muscles.

BLOOD VESSELS OF THE THIGH, LEG, AND FOOT
(Figs. 3-15, 3-16, 3-17, and 3-18, pp. 54, 56, 58, and 60)

Arteries

Femoral. Locate the artery in the femoral triangle between the femoral vein and nerve. Follow it to the dorsal side of the femur and to the popliteal fossa, where it is called the popliteal artery. (On uninjected cats it will not be practical to follow the blood vessels distal to the popliteal fossa.)

Popliteal. Within the fossa this artery divides into a posterior tibial and an anterior tibial artery.

Posterior tibial. Ramifies in the dorsal crural muscles in the cat; in the human it passes distad on the leg and into the foot region.

Anterior tibial. Pierces the interosseous membrane and continues distad, on the ventral side of the leg, to the foot.

All of the arteries give off various branches throughout their course. In the human the posterior tibial artery gives off a large peroneal branch. Arteries anastomose in the foot to form a plantar arch.

Veins

Superficial veins. The saphenous veins have been described in the section on muscles of the thigh (see p. 53).

Deep veins. These parallel the arteries and bear the same names: femoral popliteal, anterior tibial, and posterior tibial (and peroneal in the human).

There are many communicating trunks between the superficial veins, and between the superficial veins and the deep veins.

When dissection of the deep muscles of the appendages has been completed on one side of the specimen, a superficial dissection should be made on the opposite side. Separate the superficial muscles as much as possible without destroying them. Do not transect any of them.

Muscles of the Trunk

THORACIC REGION
(Figs. 3-20 and 3-21, pp. 65 and 66)

Scaleni group. Using the side of the specimen on which the pectorales and latissimus dorsi have been transected, find the scaleni muscles. This is a group of small muscles, deep to the pectorales, that arise from the cervical vertebrae and insert on the upper ribs. There are three of these: an anterior, a middle, and a posterior muscle. In the cat the middle muscle is larger than the others and extends farther caudad, crossing the serratus anterior near the ventral border of the latter. Because the scaleni assist in elevating the ribs and pulling them outward, they are synergists to other muscles of inspiration.

Medial to the scaleni is the cranial portion of the rectus abdominis and the small transversus costarum, or sternalis, superficial to it. Loosen the cranial portion of the rectus abdominis caudal to the sternalis, and transect. Reflect to the cranial border of the external abdominal oblique in order to observe the intercostal muscles.

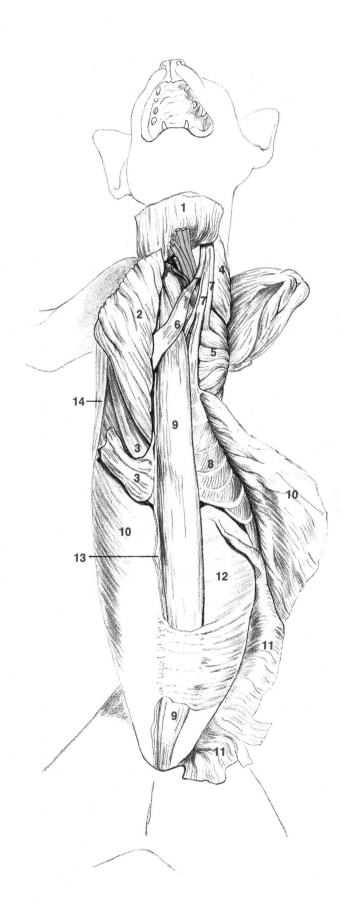

Fig. 3–20

VENTRAL VIEW
OF THE MUSCLES
OF THE ABDOMEN
AND THORAX

1. Pectoralis major
2. Pectoralis minor
3. Xiphihumeralis
4. Levator scapulae
5. Serratus anterior
6. Transversus costarum
   (sternalis)
7. Scaleni muscles
8. External intercostal
9. Rectus abdominis
10. External abdominal oblique
11. Internal abdominal oblique
12. Transversus abdominis
13. Cut edge of aponeuroses
    of abdominal muscles
14. Latissimus dorsi

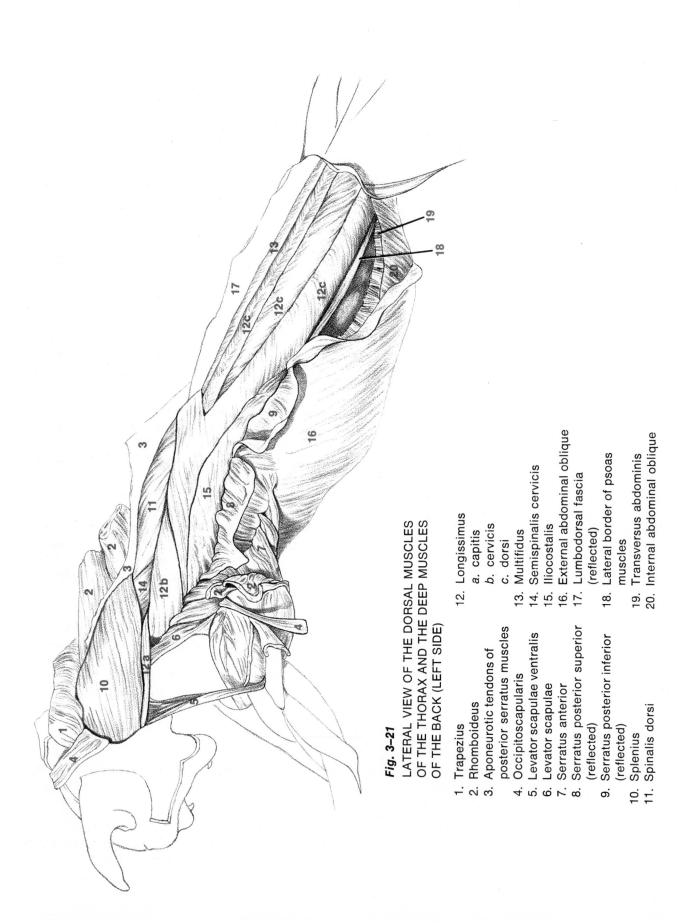

Fig. 3-21

LATERAL VIEW OF THE DORSAL MUSCLES
OF THE THORAX AND THE DEEP MUSCLES
OF THE BACK (LEFT SIDE)

1. Trapezius
2. Rhomboideus
3. Aponeurotic tendons of
   posterior serratus muscles
4. Occipitoscapularis
5. Levator scapulae ventralis
6. Levator scapulae
7. Serratus anterior
8. Serratus posterior superior
   (reflected)
9. Serratus posterior inferior
   (reflected)
10. Splenius
11. Spinalis dorsi

12. Longissimus
    a. capitis
    b. cervicis
    c. dorsi
13. Multifidus
14. Semispinalis cervicis
15. Iliocostalis
16. External abdominal oblique
17. Lumbodorsal fascia
    (reflected)
18. Lateral border of psoas
    muscles
19. Transversus abdominis
20. Internal abdominal oblique

External intercostals (Intercostales externi). There are eleven pairs in the human; twelve pairs in the cat. Note the direction of the fibers forward and downward, from the caudal border of one rib to the cranial border of the next rib caudad. Note that the external intercostals do not reach the sternum, and that the internal intercostals can be seen in the interval. The external intercostals are elevators of the ribs and are therefore muscles of inspiration.

Internal intercostals (Intercostales interni). These are immediately deep to the external intercostals, and they equal them in number. Note that the fibers course at approximately right angles to the fiber direction of the external muscles, as they pass from the cranial border of one rib to the caudal border of the next rib craniad. There is some disagreement on what the function of these muscles is, but they are probably depressors of the ribs and therefore muscles of expiration.

The space between ribs, which is occupied by the intercostal muscles, is called the **intercostal space.** Insert a probe in an intercostal space between the ventral border of the external intercostal and the underlying internal intercostal, to determine the separation.

Diaphragm. This muscle will not be observed until the internal systems are studied, and description and directions will be included in the chapter on respiratory and digestive systems. When the diaphragm contracts, the ribs are pulled outward, and the thorax is thus expanded; the diaphragm is therefore a muscle of inspiration.

Serratus posterior superior. A comparatively small muscle that can be found on the dorsal side of the trunk, deep to the rhomboideus and latissimus dorsi. In the cat this muscle extends from the upper eight or nine ribs to the cervical and thoracic spines, where it attaches by aponeurosis. In the human it is less extensive, arising from the last two cervical vertebrae and the first two or three thoracic vertebrae, and inserting on the upper ribs. When the muscle contracts it elevates the ribs and is therefore a synergist in inspiration.

Serratus posterior inferior. A small muscle extending from the last four or five ribs to the lumbar spines, where it attaches by aponeurosis. In the human, origin is on the last two thoracic vertebrae and the first two lumbar vertebrae, and insertion is on the last four ribs. This muscle assists in elongating the thorax and is a synergist in inspiration.

ABDOMINAL REGION
(Fig. 3-20, p. 65)

Note the midventral fascial line on the abdomen. This line, called the **linea alba,** is formed by a fusion of the aponeuroses of the ventrolateral abdominal muscles.

Rectus abdominis. A straight muscle extending from the pubis to the sternum on each side of the linea alba. It extends farther craniad in the cat. The origin and insertion are reversible in their actions; that is, when either end of the muscle is fixed, the opposite end can accomplish certain movements of flexion.

External abdominal oblique (Obliquus abdominis externus). A sheet of muscle in the ventral and lateral abdominal wall. It arises from the external surfaces of the lower nine or ten ribs in the cat (the lower eight in the human), with the cranial portion interdigitating with the origin of the serratus anterior. Insertion is on the iliac crest and by an aponeurosis which fuses with that of the opposite side to help form the linea alba. Note the general forward and downward direction of the muscle fibers, and compare with the external intercostal fiber direction.

Loosen the cranial border of the external abdominal oblique with a probe, and begin transecting the muscle in a caudal direction, approximately one-half inch laterad of the rectus abdominis. Carefully separate the muscle from the underlying muscles throughout its extent, transecting as the separation proceeds, and reflect.

Internal abdominal oblique (Obliquus abdominis internus). Lies immediately beneath the external abdominal oblique. This is also a sheet of fibers that arise from the iliac crest, the inguinal ligament (between the anterior superior iliac spine and the pubis), and the lumbodorsal fascia. Note that, in general, the fibers course at approximately right angles to those of the external abdominal oblique; compare with the internal intercostal fiber direction. The muscle fibers do not extend as far mediad in the cat as they do in the human. Insertion is by aponeurosis to the midline (linea alba). In the human there is also insertion on the costal cartilages of the lower ribs. The cranial two-thirds of the aponeurosis of the internal abdominal oblique splits at the lateral border of the rectus abdominis, which it encloses in a sheath (rectus sheath), and then fuses again at the midline. The caudal third of the aponeurosis does not split, and it passes ventral to the rectus abdominis.

Carefully separate the internal abdominal oblique from the underlying transversus abdominis. Because the two muscle fiber sheets and their aponeuroses are quite thin and closely applied to each other, they are difficult to separate. As the separation proceeds, transect the internal abdominal oblique in such a way that some muscle fibers will be left attached to the aponeurotic tendon. Note the relationships of the tendon with the rectus abdominis.

Transversus abdominis. The innermost one of the ventrolateral abdominal muscles, arising from the lower costal cartilages, the lumbodorsal fascia, the iliac crest, and the inguinal ligament. Insertion is by aponeurosis into the linea alba. Note the general transverse direction of the fibers.

Observe the aponeurosis of the transversus abdominis and the dorsal division of the aponeurosis of the internal abdominal oblique in their cranial two-thirds as they pass dorsal to the rectus abdominis. Note that in approximately the caudal third, all of the aponeuroses of the ventrolateral abdominal muscles pass ventrad of the rectus abdominis, so that the caudal portion of this muscle has only the transversalis fascia (a thin layer of connective tissue) and the parietal peritoneum between it and the peritoneal cavity.

Quadratus lumborum. This is a small muscle, in the cat, arising from the last two thoracic vertebrae and the last rib. It lies against the ventral surfaces of the transverse processes of the lumbar vertebrae and attaches to them; it inserts caudally on the anterior inferior iliac spine. In the human, this muscle helps to form the dorsal abdominal wall and is therefore a posterior, or dorsal, abdominal muscle. Its origin in the human is from the dorsal part of the iliac crest and adjacent ligaments, and its insertion is on the transverse processes of the upper lumbar vertebrae and the inferior border of the last rib. You will not observe this muscle until you study the deep back muscles.

Paying careful attention to muscle fiber direction, place the human abdominal muscles on Diagrams 3a and 3b. Since you have four sides, two in each diagram, to use, you will have ample room for drawing these muscles in place. Use your judgment in arranging these to best advantage for studying them.

BACK REGION
(Fig. 3-21, p. 66)

The deep muscles of the back are epaxial muscles. The other muscles studied in the cat, with the exception of the splenius, are hypaxial muscles. Epaxial muscles are supplied by the dorsal rami of spinal nerves.

Reflect the latissimus dorsi and the lumbodorsal fascia to the vertebral spines, and locate the caudal portion of the first muscle described below.

Extensor dorsi communis. A large muscle mass, on each side of the vertebral column, extending from the sacrum and ilium to the skull. This muscle mass is comparable to the **sacrospinalis,** or **erector spinae,** of the human. It is an extensor of the spine, as its name indicates. The caudal portion of the muscle arises, by a strong aponeurosis, from the iliac crest and median sacral crest, and from the lumbar vertebrae and last two thoracic vertebrae. As it extends craniad, it gives off fibers that insert on the ribs and the more cranial transverse and spinous processes. At each segmental level it receives new fibers of origin before relinquishing fibers of insertion. The intersegmental arrangement of this muscle mass provides for considerable flexibility in movement of the vertebral column.

The extensor dorsi communis is divided longitudinally into three columns. From lateral to medial these are: the **iliocostalis,** the **longissimus,** and the **spinalis.** The longissimus extends to the skull and different names are used according to the location; these are the longissimus dorsi, cervicis, and capitis. The spinalis is usually called spinalis dorsi. Some fibers of the spinalis dorsi pass cranially to join a neck muscle in the cat that is called biventer cervicis (the spinalis capitis of the human). In the human, all three columns of the sacrospinalis have three parts, which are, from lateral to medial and caudal to cranial, as follows; iliocostalis lumborum, dorsi, and cervicis; longissimus dorsi, cervicis, and capitis; spinalis dorsi, cervicis, and capitis.

The muscle columns in the thoracic and cervical regions will not be studied closely, because it is difficult to dissect them.

Multifidus. This muscle extends the length of the vertebral column, adjacent to the spinous processes. In the cat it can be observed in the lumbar region, medial to the longissimus dorsi: in the human it is covered by the sacrospinalis in this region. In the cat its cranial portions are known as the semispinalis, which is a separate muscle in the human.

Semispinalis. This muscle lies deep to the splenius. It has two parts in the cat, the semispinalis cervicis and capitis. The medial portions of the two parts form the biventer cervicis and the lateral portions form the complexus (the semispinalis capitis of the human). In the human the semispinalis has three parts: dorsi, cervicis, and capitis.

There are other small muscles in the back that assist in moving the vertebrae, but it is beyond the scope of this manual to include these.

Sever the connection of the ventrolateral abdominal muscles with the lumbodorsal fascia. Clear away the connective tissue, being *very careful* not to disturb the contents of the abdominal cavity or to destroy the parietal peritoneum bounding the peritoneal cavity. Separate and identify the lateral borders of the posterior abdominal muscles, which lie ventral to the caudal part of the extensor dorsi communis. These are, from ventral to dorsal: psoas minor, psoas major, quadratus lumborum. The iliacus will be found dorsal to the caudal portion of the psoas major, which it joins to form the iliopsoas. These muscles can be more closely observed when the contents of the abdominal cavity are studied.

Draw the human deep back muscles in place on Diagram 2*b*.

Human Skeletal Diagrams
for Muscle Placement

The following pages of diagrams are included for the use of students who are emphasizing origin, insertion, and action of human muscles.

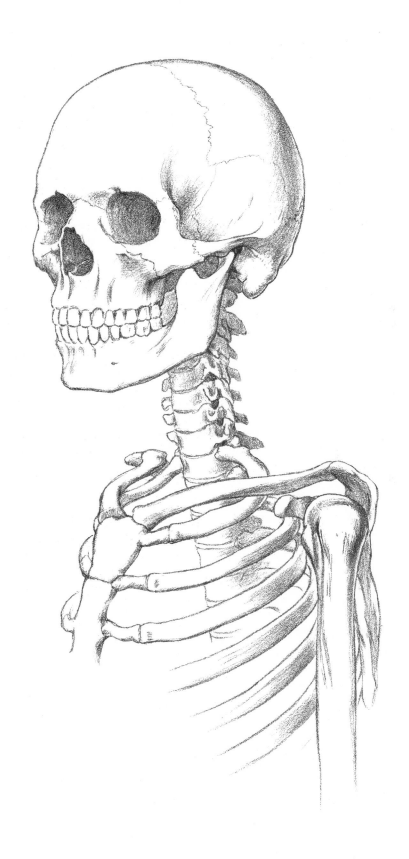

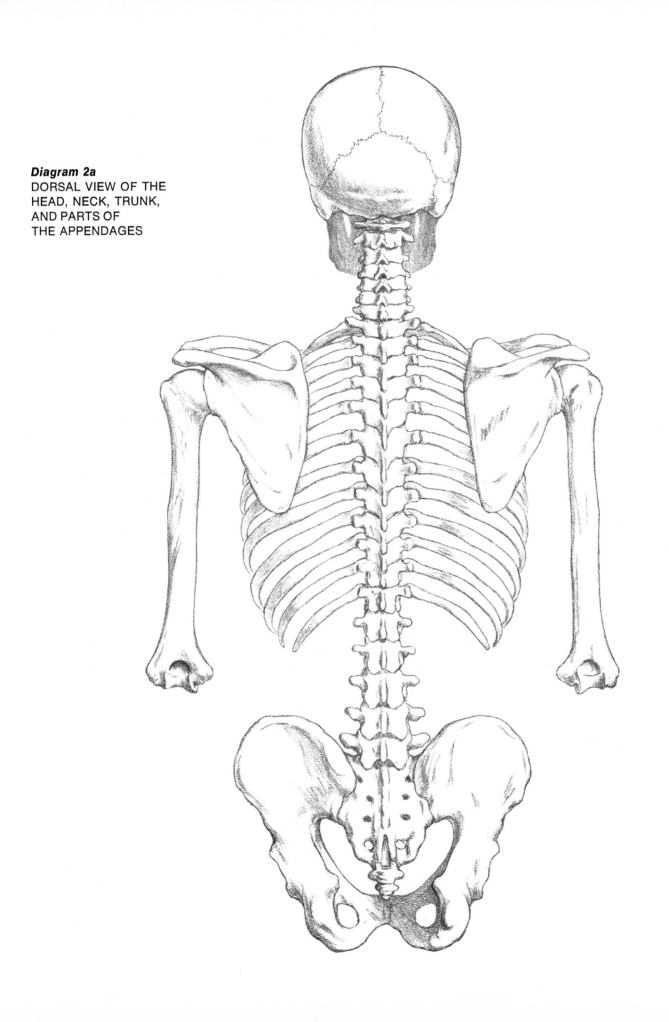

Diagram 2a
DORSAL VIEW OF THE
HEAD, NECK, TRUNK,
AND PARTS OF
THE APPENDAGES

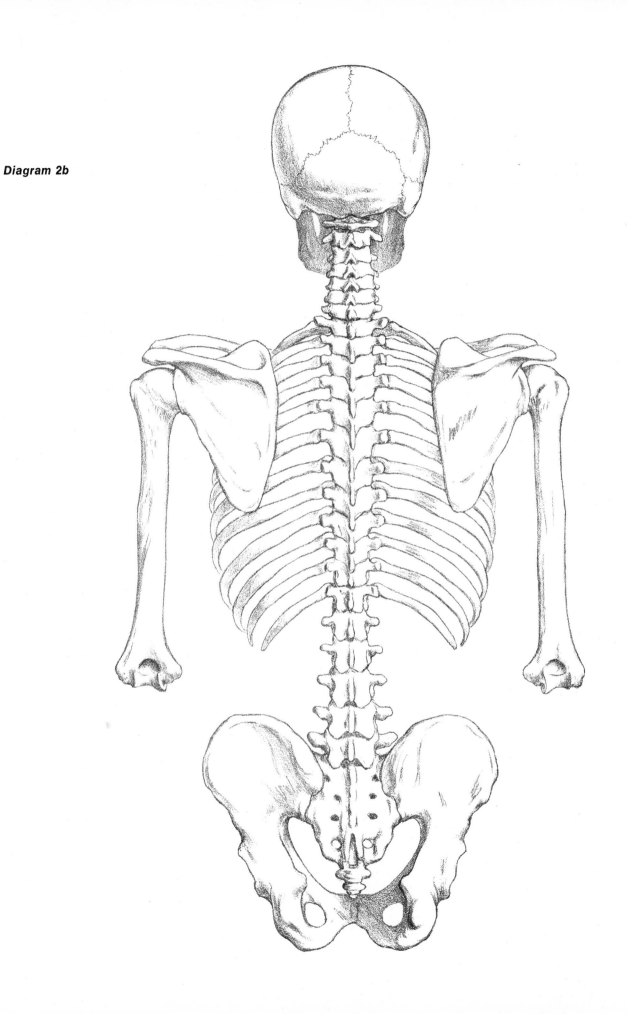

Diagram 2b

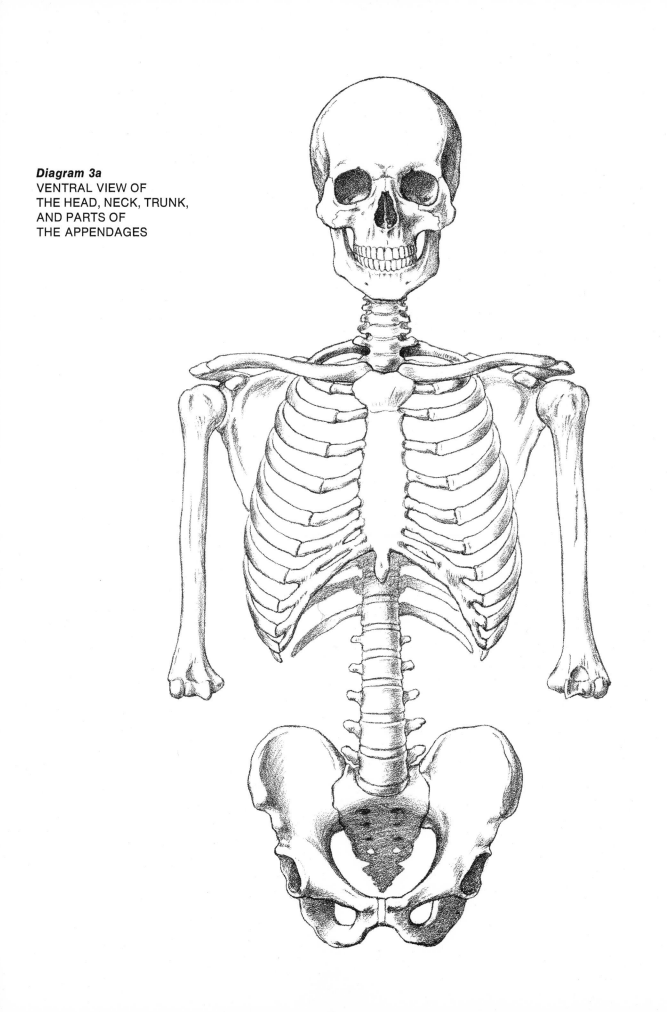

Diagram 3a
VENTRAL VIEW OF
THE HEAD, NECK, TRUNK,
AND PARTS OF
THE APPENDAGES

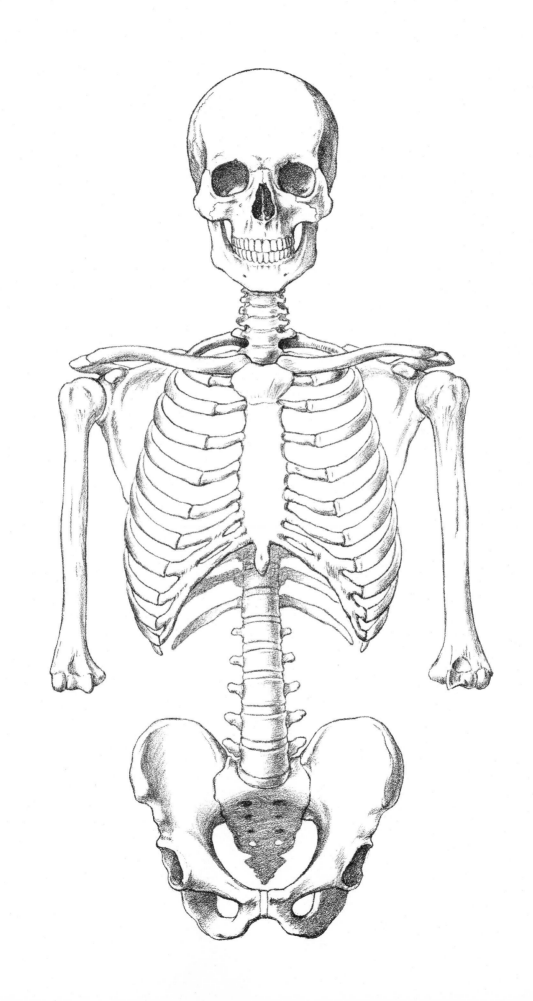

Diagram 3b

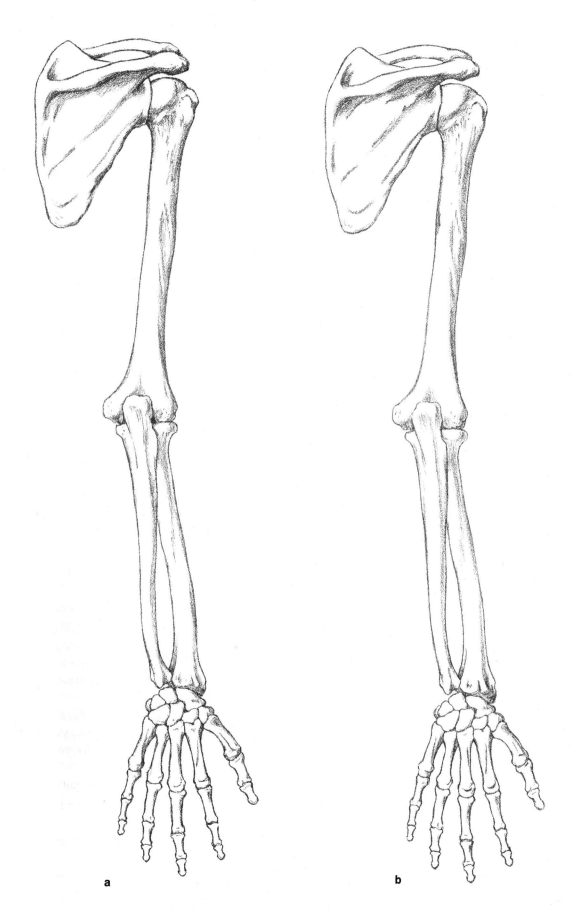

Diagram 4
DORSAL VIEW OF
THE SHOULDER,
ARM, FOREARM,
AND HAND

a

b

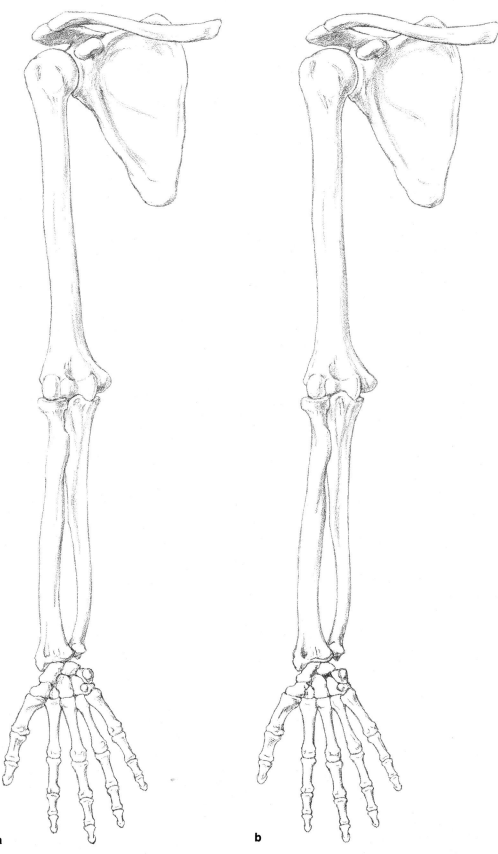

Diagram 5
VENTRAL VIEW OF
THE SHOULDER, ARM,
FOREARM, AND HAND

a

b

Diagram 6
DORSAL VIEW OF
THE PELVIC AND
THIGH REGIONS

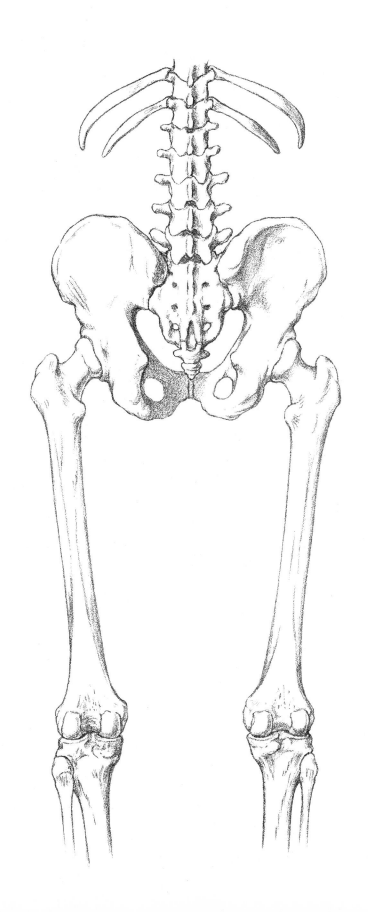

Diagram 7
VENTRAL VIEW OF
THE PELVIC AND
THIGH REGIONS

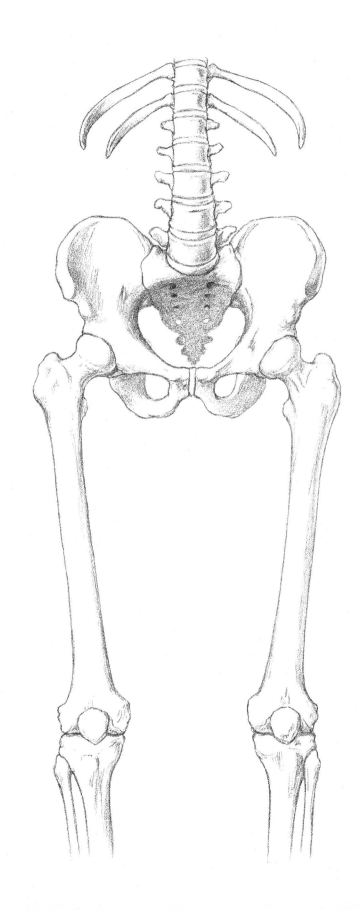

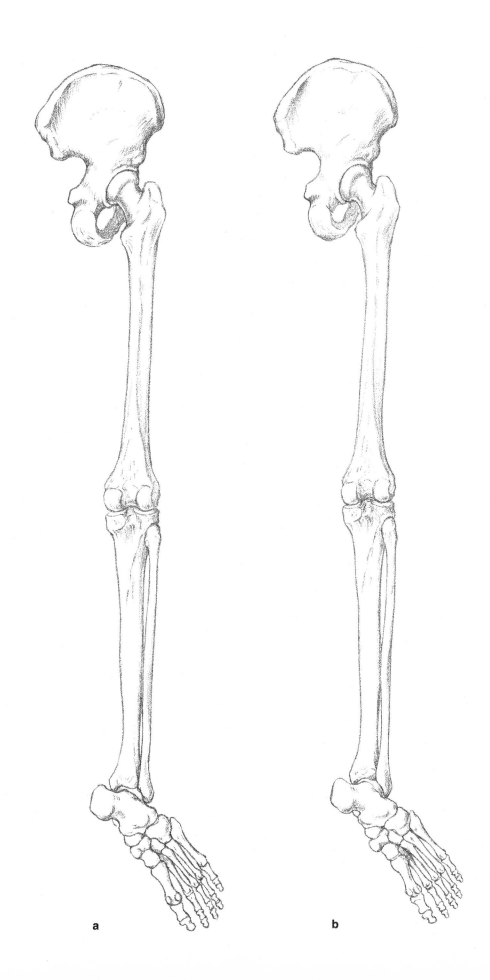

Diagram 8
DORSOLATERAL VIEW
OF HIP, THIGH, LEG,
AND FOOT,
SHOWING
THE PLANTAR SURFACE
OF THE FOOT

a

b

Diagram 9
DORSAL VIEW OF
THE LEG AND FOOT

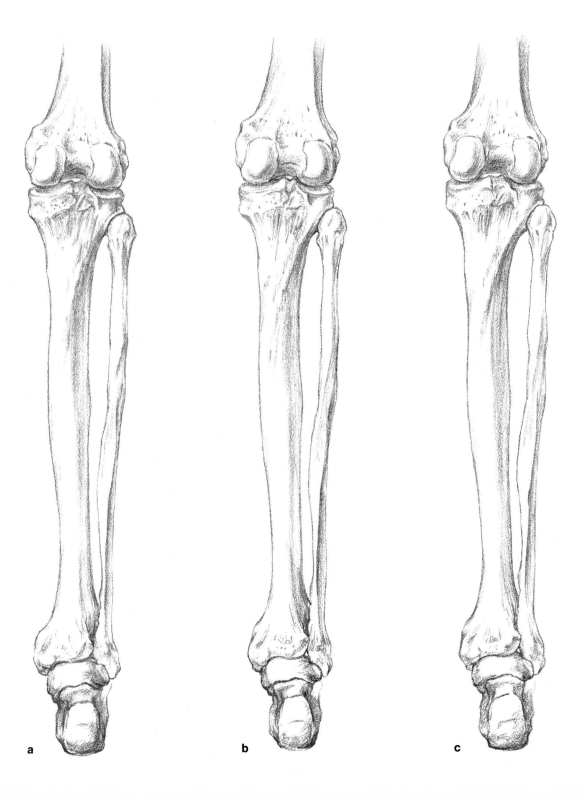

a   b   c

Diagram 10
VENTRAL VIEW OF
THE LEG AND FOOT,
SHOWING THE DORSUM
OF THE FOOT

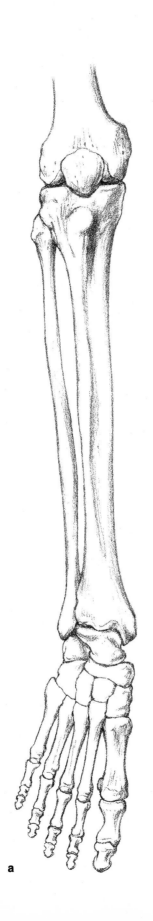

a

b

Head and Cervical Regions

The study of the respiratory and digestive systems should, perhaps, properly begin with the nasal, oral, and pharyngeal areas, which are closely related in embryonic development and remain closely related in the adult. However, in this chapter, the only structures which you will study in the head region are the salivary glands. You will observe the other structures of the head region at a later time, when you make a sagittal section of the head.

SALIVARY GLANDS
(Fig. 3-3, p. 32)

Locate the major salivary glands:

Parotid. Located ventral to the ear. Its duct crosses the masseter muscle before opening into the vestibule of the oral region.

Submandibular (submaxillary). Located ventral to the parotid gland. Its duct passes deep to the digastric muscle and opens into the floor of the oral cavity.

Sublingual. This gland is quite small. It is deep to the submandibular gland, and appears to be a part of the submandibular gland that extends ventrad. The duct courses with the submandibular duct and opens with it into the floor of the oral cavity.

SAGITTAL SECTION OF THE HEAD
(Fig. 4-1, p. 96)

This study will be postponed until a sagittal section of the head is made (see p. 126), but the observations that you should make are indicated below.

External nares, or **nostrils.** The openings into the nasal cavity.

Nasal cavity. Located above the hard palate. Note the septum that divides the cavity into two chambers, each with an external naris.

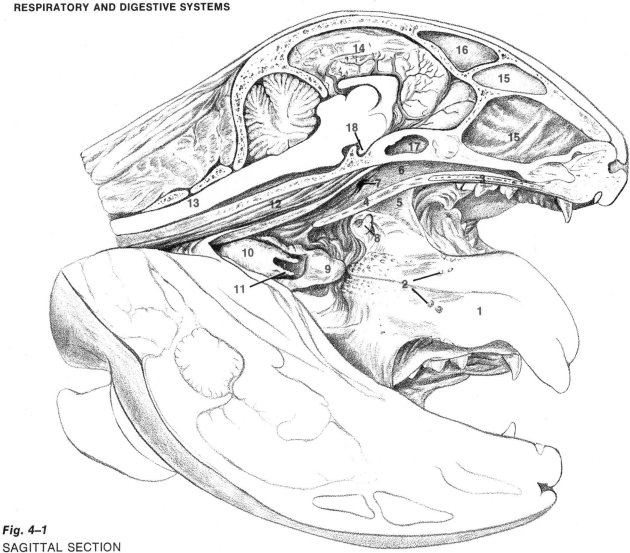

Fig. 4–1

SAGITTAL SECTION
OF THE HEAD

1. Tongue
2. Circumvallate papillae
3. Hard palate
4. Soft palate
5. Wall of oral pharynx
6. Wall of nasal pharynx
7. Opening of internal auditory tube
8. Palatine tonsil
9. Epiglottis
10. Ventral wall of esophagus at opening
11. Laryngeal cartilage
12. Pharyngeal muscles
13. Spinal cord
14. Brain
15. Nasal septum
16. Frontal sinus
17. Sphenoid sinus
18. Hypophysis

Internal nares, or **choanae.** Openings between the nasal cavity and the next portion of the respiratory pathway, the nasal pharynx.

Oral opening, or **mouth.** The opening into the vestibule of the oral region.

Vestibule. The space immediately internal to the lips and cheeks, but external to the jaws and teeth.

Tongue. The muscular structure protruding from the floor of the oral cavity. Note the papillae, particularly the circumvallate papillae on the posterior portion.

Palate. The partition, composed of hard and soft portions, between the most cranial parts of the respiratory and digestive pathways. It forms the roof of the oral cavity and oral pharynx, and the floor of the nasal cavity and nasal pharynx.

Oral cavity. The portion of the digestive pathway below the hard palate.

Oral pharynx, or **oropharynx.** The portion of the digestive pathway below the soft palate.

Nasal pharynx, or **nasopharynx.** The portion of the respiratory pathway above the soft palate.

Laryngopharynx. Immediately caudal to the oral pharynx and nasal pharynx. The respiratory and digestive pathways cross in the laryngopharynx. Note the openings of the larynx and esophagus, the latter being dorsal.

Opening of the internal auditory tube (Eustachian tube). In the lateral wall of the nasal pharynx.

Palatine, or **faucial tonsil.** In the lateral wall of the oral pharynx.

Epiglottis. The mucosa-covered cartilage ventral to the opening of the larynx. Projects upward, dorsal to the root of the tongue.

Vestibule of the larynx. Between the lateral extensions of the epiglottis and the ventricular folds.

Ventricular folds, or **"false vocal cords."** Mucosal folds between the epiglottis ventrally and small cartilages of the larynx dorsally.

Ventricle of the larynx. The part of the larynx between the ventricular folds and the vocal folds.

Vocal folds, or **"true vocal cords".** Caudal to the ventricular folds and the ventricle.

Glottis. The opening between the vocal folds.

Note the position of the brain in relation to the respiratory and digestive pathways.

LARYNX
(Figs. 3-3, 4-1, 4-2, and 4-3, pp. 32, and 96–98)

If the larynx is to be studied in detail, you will do so after the sagittal section of the head has been made (p. 126).

Certain features of the larynx were listed in the preceding section: the epiglottis, vestibule, false vocal cords, ventricle, true vocal cords, and glottis. Note the small projections in the dorsal wall of the vestibule. These are projections of the **arytenoid cartilages.**

Expose the **thyroid** and **cricoid cartilages** ventrally (using a probe and/or scalpel as needed). The cricoid cartilage is just craniad of the trachea, and the thyroid cartilage is immediately craniad of the cricoid. The hyoid bone is just craniad of the thyroid cartilage.

Note the small external muscles (Fig. 3-3) that assist in operating the laryngeal cartilages.

On one side, slightly laterad of the midline, cut the thyroid cartilage on its longitudinal axis—*just barely* through the cartilage so that the underlying muscle (the thyroarytenoid) is left intact. Carefully separate the section of thyroid cartilage from the muscle, and disarticulate it from the cricoid cartilage. Identify the following muscles that attach to the small triangular arytenoid cartilage:

Thyroarytenoid(eus). A thin, broad muscle lying lateral to the vocal folds.

Lateral cricoarytenoid(eus). A small muscle inferior to the thyroarytenoid.

Posterior cricoarytenoid(eus). From the dorsal surface of the cricoid cartilage, the fibers of this muscle run cranially and laterally to the arytenoid cartilage.

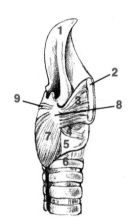

Fig. 4–2
LATERAL VIEW
OF THE LARYNX
WITH THE THYROID
CARTILAGE REMOVED
(RIGHT SIDE)

1. Epiglottis
2. Thyroid cartilage
3. Thyroarytenoid muscle
4. Lateral cricoarytenoid muscle
5. Cricoid cartilage
6. Cartilage ring of trachea
7. Posterior cricoarytenoid muscle
8. Location of arytenoid cartilage
9. Transverse arytenoid muscle

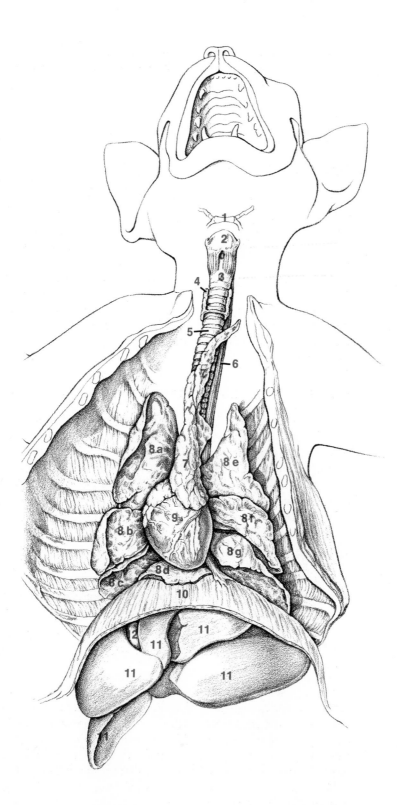

Fig. 4–3 *all*
THORACIC VISCERA

1. Hyoid bone
2. Thyroid cartilage
3. Cricoid cartilage
4. Thyroid gland
5. Trachea
6. Esophagus
7. Thymus gland
8. Lungs
   a. Right, anterior lobe
   b. Right, middle lobe
   c. Right, posterior lobe
   d. Right, mediastinal lobe
   e. Left, anterior lobe
   f. Left, middle lobe
   g. Left, posterior lobe
9. Heart
10. Diaphragm
11. Liver
12. Gall bladder

Transverse arytenoid(eus). A tiny unpaired muscle between the arytenoid cartilages. It is comparable to the transverse portion of the human arytenoid.

Thoracic Region
(Fig. 4-3, p. 98)

BODY WALL

Make an incision in the ventral thoracic body wall, about one-fourth to one-half inch to the right or left of the midline. This should pass through the costal cartilages, which can be cut with a scalpel. Extend the incision cranially to the apex of the thorax, and caudally to the cranial surface of the diaphragm.

The muscle just internal to the costal cartilages is the transversus thoracis. The thoracic body wall, from outermost layer to innermost layer, is composed of the following: integument, subcutaneous fascia, muscles and bony structures and their binding tissues, and **fascia endothoracica** (a thin layer of connective tissue). Internal to the fascia endothoracica is the parietal pleura.

PLEURA AND PLEURAL CAVITIES

The pleura is a serous membrane, composed of simple squamous epithelium and a little underlying connective tissue, which encloses the two spaces into which the lungs project. These spaces, the pleural cavities, contain a little serous fluid, but no structures. The viscera are excluded from the closed pleural cavities by the pleura. The pleura lining the thoracic body wall is called the **parietal pleura** and that covering the lungs is called the **visceral pleura.**

THORACIC CAVITY

The thoracic cavity is the potential space between the fascia endothoracica and the pleura, and within the **mediastinum**, which is between the two pleural cavities. The thoracic cavity contains the viscera of the thorax.

Pull the rib cage aside and note the attachment of the mediastinum to the ventral body wall.

LUNGS

Detach the rib cage from the diaphragm on each side by cutting along the cranial surface of the diaphragm, just under the caudal border of the rib cage. Extend these incisions far enough laterad and dorsad to expose the lungs. Note the position of the heart within the mediastinum. Using bone shears, cut the ribs dorsally, a short distance laterad of the vertebral column on each side, and pull the rib cage back. This allows a better observation of the thoracic viscera.

Note the thin smooth covering of visceral pleura over the lobes of the lungs (four lobes on the right and three on the left in the cat; three lobes on the right and two on the left in the human) and its continuity with the pleura of the mediastinum. Note the soft rubbery texture of the lung tissue (which is spongy in a living specimen).

TRACHEA

Using probe and/or scalpel as needed, extend the ventral incision cranially through the muscles and fascia of the cervical region to the larynx. Note the cartilage rings of the trachea.

Locate the small two-lobed **thyroid gland.** One lobe lies on each side of the trachea at the caudal border of the larynx. The lobes are connected across the midline, ventral to the trachea, by a thin band of glandular tissue called the **isthmus.** Because the isthmus in the cat is very small, it is usually overlooked and destroyed. The glandular tissue that is ventral to the trachea in its caudal half, and extends to the heart, is the **thymus gland.** Leave the thymus gland intact for now.

Dorsal to the arch of the aorta, the trachea divides into right and left major bronchi, but you will not be able to observe the division at this time.

ESOPHAGUS

The esophagus is dorsal to the trachea and extends a little to the left of it. Probe along the cranial half of the trachea on the left, and partially separate the trachea and esophagus. Do not disturb the thymus gland and blood vessels in the area. The esophagus can also be seen farther caudad on the left, just craniad of the diaphragm, where it lies dorsal to the heart and ventral to the aorta.

Abdominal Region
(Fig. 4-4, p. 102)

BODY WALL

Caudad of the diaphragm, make an incision just laterad of the ventral midline, through the rectus abdominis and the aponeuroses of the ventrolateral abdominal muscles, and through the **fascia transversalis** (a thin layer of connective tissue comparable to the fascia endothoracica) and **parietal peritoneum** (a serous membrane comparable to the parietal pleura) into the **peritoneal cavity.** From this ventral incision, make another incision on each side along the caudal border of the diaphragm to the dorsal body wall. This should adequately expose the abdominal viscera that project into the peritoneal cavity.

DIAPHRAGM

This is a dome-shaped, musculotendinous partition between the thorax and abdomen. Note the **central tendon,** the muscular portion, and the structures passing through the diaphragm. The opening through which the aorta passes, between **right** and **left crura** (singular: **crus**), is called the

aortic hiatus. The opening for passage of the esophagus is called the **esophageal hiatus.** The large vein passing through the **vena caval foramen** is the inferior vena cava.

The diaphragm arises from the xiphoid process of the sternum, the cartilages of the lower six ribs, and the lumbar vertebrae. The fibers insert into the central tendon.

PERITONEUM AND PERITONEAL CAVITY

The peritoneal cavity is enclosed by peritoneum, a serous membrane analogous to the pleura, and it contains a little serous fluid, but no structures. The portion of the peritoneum lining the body wall is called the **parietal peritoneum** and that covering the viscera is called the **visceral peritoneum.** The peritoneal suspensions, or supports, of the viscera have various names, which are mentioned elsewhere in this chapter.

ABDOMINAL CAVITY

The abdominal cavity is only potential space between the fascia transversalis and the peritoneum. It contains the abdominal viscera. Note the relationships of the abdominal viscera when you first open the peritoneal cavity. Note the apron, with fat deposits, covering the intestinal portion of the digestive viscera.

LIVER

Note the lobes and ligaments of the liver. The **falciform ligament** divides the liver essentially into right and left halves. The **round ligament,** which represents the vestige of an embryonic blood vessel, is a fibrous strand within the free border of the falciform ligament. Note the suspensory ligaments attaching the liver to the diaphragm. You will find the **gall bladder,** partially surrounded, in the right median lobe of the liver. Ducts of the liver and gall bladder run in the **lesser omentum,** which is described elsewhere.

STOMACH

Lift the lobes of the liver and note the position of the stomach. Observe the esophagus passing through the diaphragm to join the stomach at its **cardiac** end. Note the **pyloric** portion of the stomach and the constriction that marks the union of the stomach with the intestine, and the location of the pyloric valve. The **pyloric sphincter muscle,** which produces the valve action, is a thickening of the circular muscle layer at the caudal end of the stomach. Note the **greater** and **lesser curvatures** of the stomach, the greater on the left, and the lesser on the right.

SPLEEN

The spleen, which is not a part of the digestive system, will be found along the greater curvature side of the stomach, to which it is attached by the **gastrosplenic ligament.**

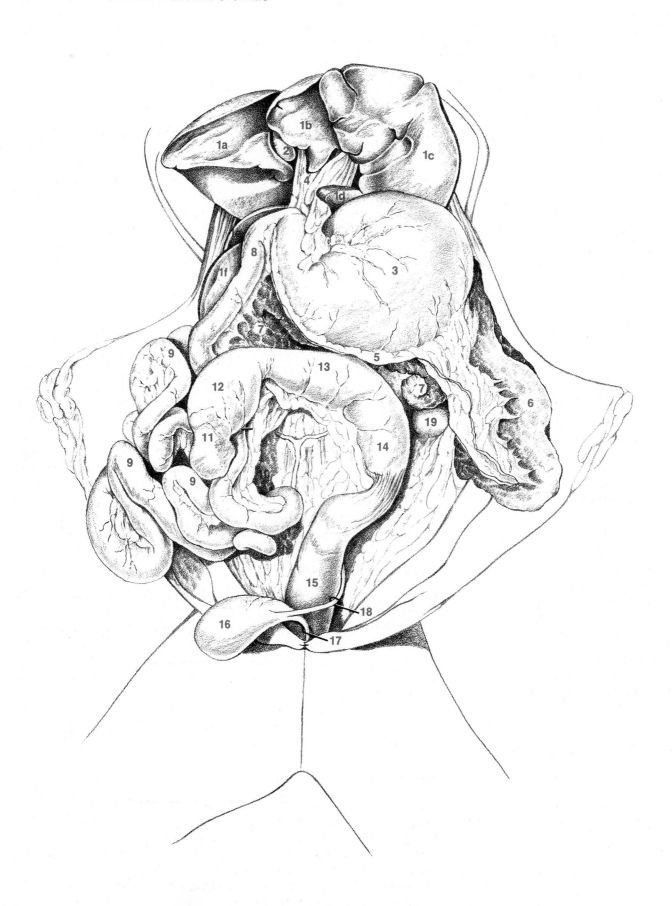

INTESTINE

The intestinal portion of the digestive tube is divided into a **small intestine** and a **large intestine.** The small intestine consists of the **duodenum** (the cranial and shortest portion), **jejunum,** and **ileum.** The ileum joins the large intestine at the junction of the cecum and the colon. The large intestine consists of the **cecum, colon** (which has ascending, transverse, and descending portions), **rectum,** and **anal canal.** In the human there is also an S-shaped portion of the colon, between the descending colon and rectum, which is called the sigmoid colon. The digestive tube, which begins cranially with the oral opening, or mouth, ends caudally with the anal opening or **anus.**

LESSER OMENTUM

This is a double layer of peritoneum extending from the lesser curvature of the stomach and from the duodenum to the liver. The portion between the stomach and the liver is the **hepatogastric ligament**; that between the duodenum and the liver is the **hepatoduodenal ligament.** The lesser omentum contains blood vessels and the bile ducts between its layers. The **common bile duct** is enclosed in the free edge and can be observed joining the duodenum.

Dorsal to the free border of the lesser omentum, an opening, the **epiploic foramen,** or **foramen of Winslow,** leads into the **lesser peritoneal cavity,** which is dorsal to the lesser omentum and stomach and extends into the omental bursa (which is described in the following paragraph). The portion of the peritoneal cavity that the abdominal viscera project into is called the **greater peritoneal cavity.**

GREATER OMENTUM

Note the double layer of peritoneum that attaches the stomach, by its greater curvature, to the dorsal body wall. This is the greater omentum, which encloses the spleen and part of the pancreas between its layers. The apron-like, double-layered sac extending caudad, ventral to the intestine, is called the **omental bursa** or **lesser peritoneal sac,** and it is a part of the greater omentum, as is the gastrosplenic ligament between the stomach and spleen.

Lift the omental bursa and spread it out. Note the fat deposits. Carefully separate the dorsal and ventral walls, and note the space (part of the lesser peritoneal cavity) within.

MESENTERY

The mesentery suspends the small and large intestine from the dorsal body wall. The part that suspends the small intestine is called the **mesentery proper.** Note the great difference in length between the body wall attachment and the small intestine. The part of the mesentery suspending the colon is called the **mesocolon** and that suspending the rectum is called the **mesorectum.** Note the lymph nodes in the mesentery, and the blood vessels coursing between the layers.

Fig. 4–4 *all*

ABDOMINAL VISCERA

1. Liver
   a. Right median lobe
   b. Left median lobe
   c. Left lateral lobe
   d. Caudate lobe
   e. Right lateral lobe, anterior (cranial) part
   f. Right lateral lobe, posterior (caudal) part
2. Gall bladder
3. Stomach
4. Lesser omentum
5. Cut edge of omental bursa (greater omentum)
6. Spleen
7. Pancreas
8. Duodenum
9. Jejunum and ileum
10. Ileocecal junction
11. Cecum
12. Ascending colon
13. Transverse colon
14. Descending colon
15. Rectum
16. Urinary bladder
17. Urethra
18. Ureter
19. Kidney (retro-peritoneal)

PANCREAS

Reflect the omental bursa craniad, and find the pancreas. You can cut the bursa near its attachment to the stomach and spleen, but leave enough attached so that the blood vessels running near the spleen and the greater curvature of the stomach will not be destroyed.

The **head** and **neck** of the pancreas lie in the curve of the duodenum, within the mesentery proper; the **body** extends to the left, between layers of the greater omentum, dorsal to the stomach, and the **tail** extends as far to the left as the spleen. (In the human, processes of development cause the pancreas to assume a retroperitoneal position.) The **pancreatic duct** joins the common bile duct within the wall of the duodenum. Take care not to destroy the blood vessels when you are attempting to locate the duct. There may be an accessory pancreatic duct opening independently into the duodenum.

The urogenital system includes both the urinary (excretory) and reproductive systems. These systems are closely associated in embryonic development and maintain a close relationship in the adult.

Urinary System
(Figs. 5-1 and 5-2, pp. 107 and 108)

Locate the **kidneys** lying against the dorsal body wall caudad of the diaphragm. They are not suspended by peritoneum, but are retroperitoneal. In the cat the right kidney lies slightly more craniad than the left one; the positions are reversed in the human. In the cat an adrenal (or suprarenal) gland is located at the cranial end of each kidney, but it is separate and lies slightly mediad of it; in the human the gland actually "caps" the tip of the kidney.

Find the **ureters**, which lead to the **urinary bladder**, coursing along the ventral surface of the psoas muscles. The ureters are also retroperitoneal, and you must break and pull aside the peritoneum in order to find them. The urinary bladder is supported, close to the ventral body wall, by a ventral suspensory ligament and by lateral ligaments. The lateral ligaments usually contain fat. The duct leading from the bladder is the **urethra.** The female urethra opens into a vestibule, serving both the urinary and the reproductive systems, which in turn opens to the exterior. The urinary bladder of the male cat (but not of the human) has a "neck," which extends from the bladder to the prostate gland, where it becomes continuous with the prostatic portion of the urethra. You will examine the urethrae in more detail when you study the reproductive system.

On the medial side of the kidney note the **hilus**, the point through which the blood vessels and the ureter enter, or leave, the kidney. Make a longitudinal section of the kidney that will divide it into ventral and dorsal halves. Observe the **cortex**, the **medulla**, and the **pyramid**, which has been formed by the coalescence of collecting tubules. The pyramid projects into the **renal pelvis,** which is the expanded portion of the ureter within the **renal sinus** (the space within the hilus). There are several pyramids in the human kidney.

Female Reproductive System
(Fig. 5-1, p. 107)

Locate the **uterus** dorsal to the urinary bladder and urethra. Note the **body** of the uterus and the **cornua**, or horns, extending laterad and craniad from the body. (The human has no uterine horns.) On each side the horn is continuous with the **uterine tube**, or **oviduct**, which curves around the **ovary**. The uterine tube opens into the peritoneal cavity through its **ostium**. Note the position of the ovary caudad of the kidney. Identify the ligaments: the **broad ligament** of the uterus (peritoneum), the **round ligament** of the uterus (the narrow fibrous band or cord enclosed between the layers of broad ligament), the **mesovarium** (the cranial extension of the broad ligament), and the **ovarian ligament** (a short, thick cord within the broad ligament, extending from ovary to uterus).

Cut through the pubic symphysis and bend the thighs dorsad to loosen the attachments. It may be necessary to cut away some of the pubic bone with bone shears. Separate binding tissues to expose the urethra, the body of the uterus, and the **vagina**, which is dorsal to the urethra. Follow the course of the urethra and vagina caudally to the site at which both open into the **vestibule**, which in turn opens to the exterior. The position of the **cervix** of the uterus can sometimes be determined externally: it is a knot of tissue (dorsal to the urethra) that is harder than the body of the uterus just craniad, and the vagina just caudad.

To observe the internal appearance of the vagina and uterus, make an incision in their ventral walls and reflect. It is easy to locate the cervix internally.

Note the dorsal position of the rectum and anal opening.

Male Reproductive System
(Fig. 5-2, p. 108)

Caudad of the pelvic region, locate the **penis**, which contains a part of the urethra, and the **scrotum**, an internally divided, integumentary sac which encloses the **testes**. Find the two **spermatic cords**, each immediately laterad of the ventral midline (at the pubic symphysis). Each cord contains nerves, and blood and lymphatic vessels, as well as a sperm duct, the **ductus** or **vas deferens**.

Using the probe to loosen attachments a little, follow a spermatic cord caudad to the scrotum, snipping connective tissues only as necessary. Make an incision in the ventral wall of the scrotal sac in order to expose the testis. To observe the testis and the **epididymis** you will have to slit and reflect the outer fascial covering of these structures. The highly convoluted **duct of the epididymis** (within the epididymis) is continuous with the ductus deferens, which is also convoluted for some distance.

Follow the ductus deferens and other structures of the spermatic cord craniad to the **external inquinal ring**. Find the ductus deferens and internal spermatic blood vessels passing through the **internal inguinal ring**. The short channel between the inguinal rings is called the **inguinal canal**. Follow the ductus deferens as it passes ventral to the ureter, then

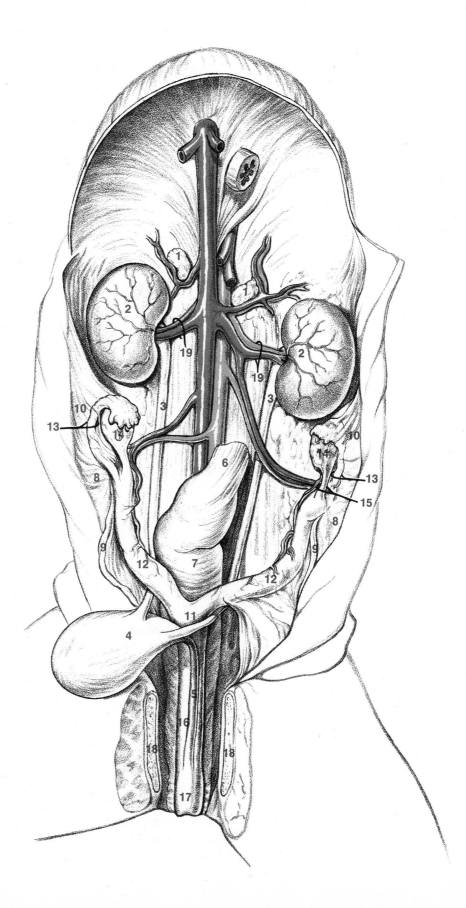

Fig. 5–1

FEMALE
UROGENITAL SYSTEM

1. Adrenal gland
2. Kidney
3. Ureter
4. Urinary bladder
5. Urethra
6. Descending colon
7. Rectum
8. Broad ligament
9. Round ligament
10. Mesovarium
11. Body of uterus
12. Horn of uterus
13. Uterine tube
14. Ovary
15. Ovarian ligament
16. Vagina
17. Vestibule
18. Pubic bone
19. Renal blood vessels

Fig. 5–2

MALE
UROGENITAL SYSTEM

1. Adrenal gland
2. Kidney
3. Ureter
4. Urinary bladder
5. Neck of bladder
6. Prostate gland
7. Pubic bone
8. Urethra
9. Bulbourethral gland
10. Crus of penis
11. Ischiocavernosus muscle
12. Penis
13. Spermatic cord with fascial
    covering
14. Testis and epididymis
    with fascial covering
15. Scrotal integument
16. Epididymis
17. Testis
18. Ductus deferens
19. Location of inguinal canal
20. Internal spermatic blood
    vessels
21. Rectum
22. Descending colon
23. Renal blood vessels

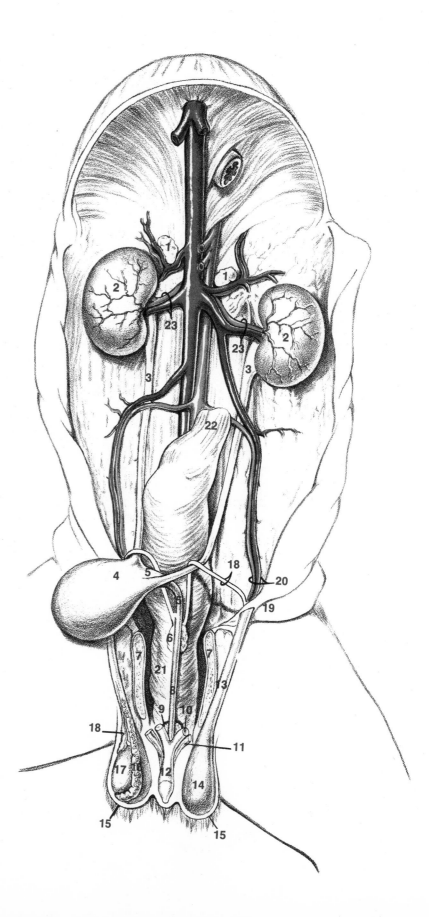

medially to a position dorsal to the urinary bladder. The ductus deferentes join the prostatic portion of the urethra.

Cut through the pubic symphysis and bend the thighs dorsad, to loosen the attachments. Carefully break binding tissues to expose the neck of the urinary bladder and the **prostate gland** (small in the cat). (In the human the prostate gland is immediately caudad of the bladder, which has no neck.)

The male urethra has three portions: prostatic, membranous, and cavernous. The **prostatic urethra** is the portion passing through the prostate gland, which empties its secretions into the urethra. Note that the ductus deferentes approach each other at the midline and run close together within a fold of peritoneum to reach the urethra within the substance of the prostate. (In the human the ductus deferens is joined by a seminal vesicle and then continues through the prostate, as the ejaculator duct, to join the prostatic urethra.) The **membranous urethra** is the portion between the prostate and the **bulbourethral glands** (Cowper's glands). Locate the bulbourethral gland on one side. To do this, cut the **crus** of the penis (the proximal end of the cavernous body) and a small muscle (the ischiocavernosum) from their attachments on the ischium. The gland will be found dorsal to these structures. Distal to the point at which the ducts of the bulbourethral glands join the urethra is the **cavernous urethra,** which courses through the penis to open to the exterior through the terminal portion of the penis (the glans penis.)

The penis is composed of three cylindrical cavernous masses of tissue with an integumentary covering. The two dorsal and larger masses are called the **cavernous bodies** and the smaller ventral mass, which contains the cavernous urethra, is called the **spongy body.**

Note that the male urethra has a dual function: reproductive and excretory.

Each student should study both female and male specimens.

Vessels Craniad of the Diaphragm

HEART
(Figs. 4-3 and 6-1, pp. 98 and 113)

Note the position of the heart within the mediastinum. Carefully clear away any thymus tissue and fat that may obscure the heart and its attached vessels. (The nerve ventral to the root of the lung on each side is the **phrenic**, which passes to the diaphragm. On the left side the nerve passing ventral to the arch of the aorta and then dorsal to the root of the lung is the **vagus**. On the right, you can find this nerve along the side of the trachea and then passing dorsal to the root of the lung.)

Slip a probe under the outer wall, the **parietal pericardium**, of the **pericardial sac** that surrounds the heart. Slit this outer wall in a craniocaudal direction and reflect from the heart. Note the attachment of the fibrous outer layer of parietal pericardium onto the blood vessels that join the heart, and the continuity of the inner serous layer with the serous inner wall of the sac. The inner wall is the **visceral pericardium** (epicardium), which invests the heart so closely that it is virtually impossible to separate it from the underlying **myocardium** (heart muscle). The space enclosed by the inner and outer walls of the pericardial sac is the **pericardial cavity.**

Note the right and left **atria** (auricles) and the right and left **ventricles** of the heart; note also the attached blood vessels: the **superior** and **inferior venae cavae** joining the right atrium, the **pulmonary aorta**, or artery, arising from the right ventricle, the **pulmonary veins** joining the left atrium, and the ascending portion of the **aorta** arising from the left ventricle. Note the apex of the heart directed caudally, and the slight groove between the right and left ventricles on the ventral side. **Coronary** blood vessels course in the groove and in sulci between the atria and ventricles.

Make incisions in the heart as indicated in the diagrams on the next page. Incision 1 passes through the ventral wall of the right ventricle into the base of the pulmonary aorta; incision 2 passes through the wall of the right atrium and through the opening of the superior vena cava; incision 3 passes through the opening of the inferior vena cava. Clean out any co-

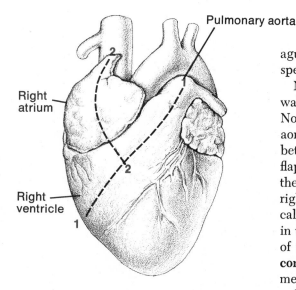

Pulmonary aorta

Right atrium

Right ventricle

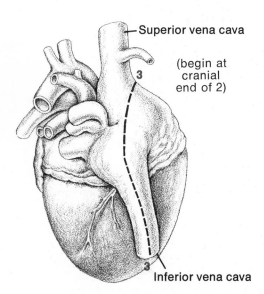

Superior vena cava

(begin at cranial end of 2)

Inferior vena cava

**THE INCISIONS
TO MAKE WHEN
OPENING THE HEART**

agulated blood that may be present in the heart chambers of uninjected specimens. Carefully remove the latex from injected specimens.

Note the difference in thickness and structure of ventricular and atrial walls, and the endothelial lining, called **endocardium**, of the chambers. Note the **semilunar valve flaps** and **sinuses** at the base of the pulmonary aorta. The valve flap is a fold of endothelium with a little connective tissue between layers of the fold; the semilunar sinus is the space between the flap and the wall of the vessel. Note the **atrioventricular valve flaps** and the attachments to the **papillary muscle** by fine **chordae tendineae**. The right atrioventricular valve is called the **tricuspid valve**; the left one is called the **bicuspid valve**, or the **mitral valve**. Note the slight depression in the interatrial septum. This is the **fossa ovalis**, which marks the location of the foramen ovale (a fetal opening between atria). The opening of the **coronary sinus**, which returns blood from the heart wall, can be found medial to the opening of the inferior vena cava. **Pulmonary arteries** and **veins** can be observed dorsally.

Because structures on the left are basically the same, except that the wall of the left ventricle is much thicker than that of the right, opening the left side of the heart is optional.

The following directions may vary at the instructor's discretion. *After the study of the nervous system has been completed,* remove the heart from *one specimen at each table,* so that the dorsal side can be observed. Sever vessels in such a way that part of each is left attached to the heart. Cut the aorta at the junction of the ascending portion and the arch, so that the vagus nerve will not be disturbed. When cutting the pulmonary vessels, take care that the vagus and phrenic nerves are not destroyed. Observe the division of the trachea into the right and left major bronchi, as well as the pulmonary vessels between the heart and lungs.

ARTERIES
(Fig. 6-2, p. 114)

Locate as many of the following arteries as possible, using a probe to separate the vessels from surrounding tissues.

Pulmonary aorta, or **artery**. Carries deoxygenated blood to the lungs via its divisions, the right and left pulmonary arteries.

Aorta. Carries oxygenated blood for all parts of the body. Note ascending, transverse (arch), and descending portions. Note the relationship to the trachea, bronchus, and esophagus. Identify the following branches:

  Coronary. Paired; from the base of the ascending aorta. The coronary arteries supply blood to the heart wall.

  Brachiocephalic (innominate). Unpaired; from the arch. Supplies the head, neck, and right pectoral appendage.

  Left subclavian. From the arch. Supplies the left pectoral appendage.

  Parietal. The intercostal, subcostal, and phrenic branches of the descending aorta. They supply body wall structures and the diaphragm.

  Visceral. The bronchial, esophageal, and pericardial branches of the descending aorta. (Some are paired, some are not.)

a

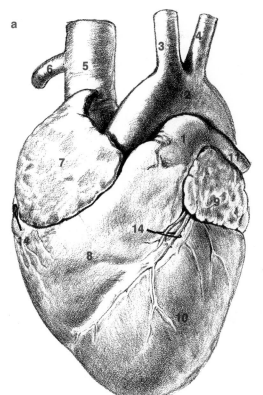

b

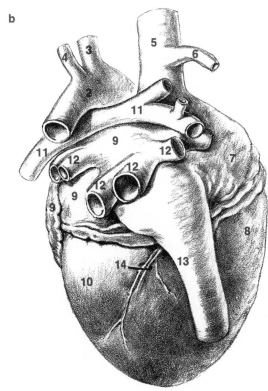

Fig. 6–1

THE HEART AND
ATTACHED VESSELS:
(a) ventral view;
(b) dorsal view

1. Pulmonary aorta
2. Arch of aorta
3. Brachiocephalic artery
4. Left subclavian artery
5. Superior vena cava
6. Azygos vein
7. Right atrium
8. Right ventricle
9. Left atrium
10. Left ventricle
11. Pulmonary arteries
12. Pulmonary veins
13. Inferior vena cava
14. Coronary blood vessels

In the human there are three branches from the arch of the aorta: the brachiocephalic, the left common carotid, and the left subclavian.

Brachiocephalic. This branch of the aorta is described in the preceding list. It has the following branches.

> **Right subclavian.** Supplies the right pectoral appendage.
> **Right common carotid.** The carotids are described in the following paragraphs.
> **Left common carotid.** (In the cat.)

In the human, the branches of the brachiocephalic artery are typically the right common carotid and right subclavian only, with the left common carotid arising directly from the arch of the aorta.

The common carotid artery will be found laterad of the trachea, on each side, in a connective tissue sheath (the carotid sheath), along with the vagus nerve, the cervical portion of the sympathetic trunk, and the internal jugular vein.

The carotid arteries carry a blood supply for the head region. Near the base of the skull each common carotid divides into an **external carotid**, which supplies primarily the head structures outside the cranial cavity, and an **internal carotid**, which supplies primarily the structures within the cranial cavity. The internal carotid of the cat, which is not entirely analogous to that of the human, is very small and may be absent. If present it can be located on injected specimens.

Common carotid. The branches that may be located on injected cats are the following:

Various small branches to muscles
Superior thyroid
Inferior thyroid. Because this is very tiny, it is not usually observed.

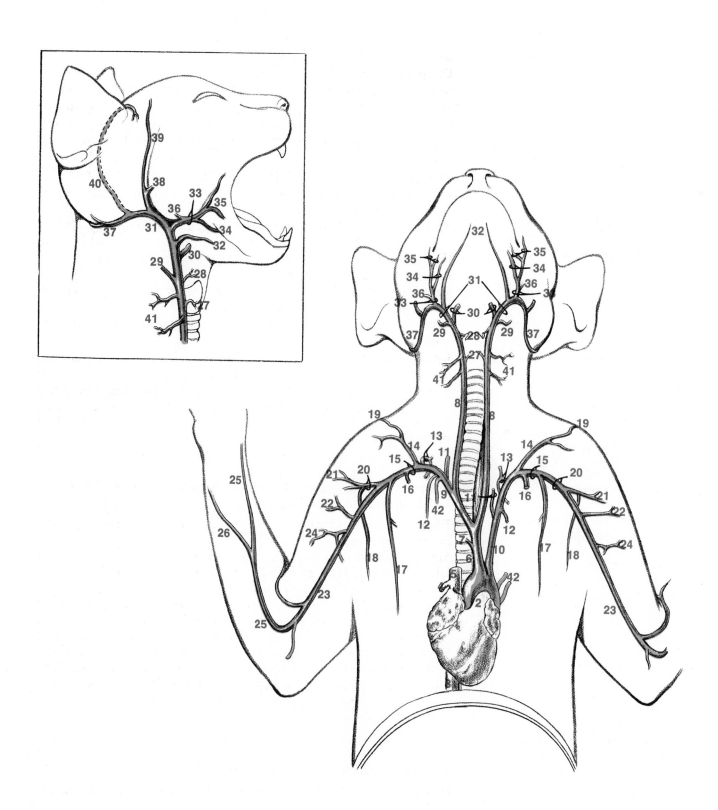

Superior laryngeal. Supplies the larynx.
Occipital. Supplies the occipital region.
External and internal carotids. Described in the preceding paragraphs.

External carotid. Locate the external jugular, anterior facial, posterior facial, and transverse veins (see p. 117 and Fig. 6-2, p. 116). Do not destroy these superficial veins, but loosen the submandibular and sublingual glands (do not cut the ducts). Loosen the parotid gland, cut the duct, and reflect dorsad. Locate the branches of the external carotid:

Lingual. Runs along the ventral border of the digastric muscle, accompanied by the hypoglossal nerve.
Branch to the submandibular and sublingual glands. In some specimens this is a branch of the external maxillary.
External maxillary. Passes deep to the digastric muscle.
Posterior auricular. Passes dorsal to the auricula.
Superficial temporal. Ventral to the ear. It branches off deep to the parotid and extends into the temporal region.
Internal maxillary. A continuation of the external carotid deep to the masseter muscle.

Subclavian. Supplies the head, neck, and thoracic wall, as well as the pectoral appendage. Its branches are the following:
Vertebral. Turns dorsad and enters the transverse foramen of the sixth cervical vertebra and passes through the successive transverse foramina en route to the brain.
Costocervical trunk. Supplies costal and cervical regions.
Thyrocervical trunk. Continues in the shoulder region as the transverse scapular (suprascapular) artery.
Internal mammary (internal thoracic). Supplies primarily the ventral thoracic wall. It has many intercostal branches.

Axillary. The continuation of the subclavian artery into the axilla. Its branches are these:
Anterior thoracic. Supplies the pectoralis muscles.
Long thoracic (lateral thoracic). Supplies the pectorales, the serratus anterior, and the latissimus dorsi.
Subscapular. It has two major branches: the **posterior humeral circumflex** to shoulder muscles and dorsal muscles of the arm; and the **thoracodorsal** to the teres major and the latissimus dorsi. Other branches that may be given off in some specimens are the **anterior humeral circumflex** and **deep brachial,** but these are usually branches of the brachial artery. The main trunk of the subscapular continues on to supply the subscapularis and other muscles on the dorsal side of the scapula.

Major Arteries of the Pectoral Extremity

Brachial. A continuation of the axillary artery into the arm region. Its branches are:
Anterior humeral circumflex. Extends to the biceps brachii and the head of the humerus.

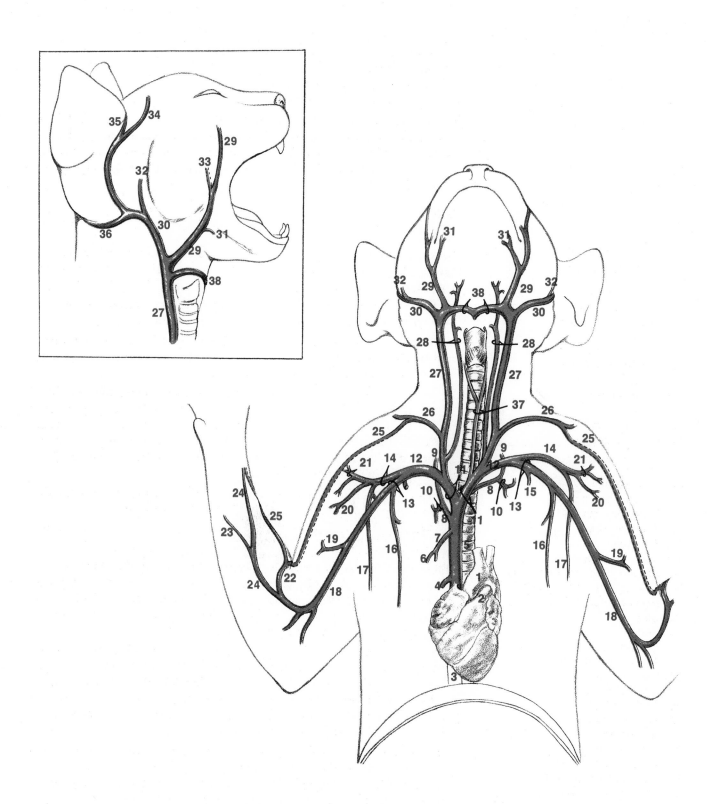

Deep brachial. Extends to the triceps brachii, epitrochlearis, and latissimus dorsi.

Various other arteries to muscles

Arteries to the elbow region

Radial. The brachial artery continues distal to the elbow as the radial. The radial has various branches, including the small **ulnar** in the cat. In the human the brachial divides into radial and ulnar arteries immediately distal to the bend of the elbow, the ulnar usually being the larger.

VEINS
(Fig. 6-3, p. 116)

Veins, in general, course parallel to the arteries of the same name, but there are some notable exceptions. Locate as many of the following veins as possible.

Inferior vena cava. Unpaired vessel joining the right atrium. It returns all blood to the heart from areas caudad of the diaphragm.

Superior vena cava. Unpaired vessel joining the right atrium. It returns all blood to the heart from areas craniad of the diaphragm, except that from the wall of the heart itself.

Azygos. Unpaired vessel located on the right side along the vertebral column. It joins the superior vena cava just before the latter joins the right atrium. The azygos collects blood from the thoracic wall from intercostal tributaries, and some from the diaphragm. It receives **esophageal veins** from the esophagus and **bronchial veins** from the lungs. The human has accessory azygos veins.

Pulmonary. Returns oxygenated blood from the lungs to the left atrium. There may be two or more of these; the human has four.

Brachiocephalic (innominate). The right and left veins join to form the superior vena cava.

Subclavian. Joins the external jugular vein to form the brachiocephalic. The subclavian is a continuation of the axillary vein; therefore it returns blood from the pectoral extremity. In the human, the subclavian and internal jugular join to form the brachiocephalic, with the external jugular being a tributary to the subclavian.

External jugular. Formed by the union of the **anterior** and **posterior facial veins**, which have tributaries that return blood distributed by the branches of the external carotid artery; therefore the external jugular returns blood primarily from structures of the head region outside the cranial cavity. The two veins communicate near the point of formation by a large **transverse vein**, which passes ventral to the neck.

Internal jugular. Formed by a union of veins that are primarily from structures within the cranial cavity. It may join either the external jugular or the brachiocephalic in the cat. In the cat the internal jugular is usually quite small, and the external jugular quite large; in the human the internal jugular is the larger.

Vertebral. Accompanies the vertebral artery. In the cat the vertebral vein

Fig. 6–3
VEINS CRANIAD OF THE DIAPHRAGM

1. Arch of aorta
2. Pulmonary aorta
3. Inferior vena cava
4. Azygos
5. Superior vena cava
6. Internal mammary
7. Sternal
8. Costovertebral
9. Vertebral
10. Costocervical
11. Brachiocephalic
12. Subclavian
13. Axillary
14. Subscapular
15. Anterior thoracic
16. Long thoracic
17. Thoracodorsal
18. Brachial
19. Deep brachial
20. Anterior humeral circumflex
21. Posterior humeral circumflex
22. Median cubital
23. Ulnar
24. Radial
25. Cephalic
26. Transverse scapular
27. External jugular
28. Internal jugular
29. Anterior facial
30. Posterior facial
31. Submental
32. Internal maxillary
33. Deep facial
34. Superficial temporal
35. Anterior auricular
36. Posterior auricular
37. Inferior thyroid
38. Transverse

joins the costocervical to form a common trunk, which in turn joins the brachiocephalic on the left side, and on the right side may join either the brachiocephalic or the superior vena cava. In the human the vertebral vein joins the brachiocephalic directly.

Internal mammary (**internal thoracic**). In the cat the right and left veins unite to form the **sternal vein,** a tributary to the superior vena cava. In the human the internal mammary vein joins the brachiocephalic.

Transverse scapular (**suprascapular**). Joins the external jugular. In the cat it is joined by the cephalic vein from the extremity.

Axillary. Located in the axilla. This is a continuation of the **brachial vein** from the arm. In the human, the brachial and **basilic** veins unite to form the axillary. The tributaries to the axillary vein correspond, in general, to the branches of the axillary artery.

Major Veins of the Pectoral Extremity

Deep veins. These are the brachial vein and its tributaries, which parallel, in general, the brachial artery and its branches.

Superficial veins

  Cephalic vein (**Vena cephalica**). You previously observed this vein on the dorsal side of the pectoral extremity when you removed the skin from the cat and when you studied the pectoral appendage. It joins the transverse scapular in the cat, but the axillary vein in the human.

  Median cubital (**Vena mediana cubiti**). You observed this vein, too, when you studied the pectoral appendage. It is a communicating vein between the cephalic and brachial veins in the cat, and between the cephalic and basilic veins in the human. (The basilic vein of the human is located on the medial side of the extremity. It joins the brachial, as noted above, in formation of the axillary vein.)

LYMPHATIC VESSELS

Locate the **thoracic duct** craniad of the arch of the aorta, immediately laterad of the esophagus on the left, and follow it to its union with the venous system at the junction of the external jugular and subclavian veins. On injected specimens it may be possible to locate the right and left **lymphatic trunks,** which return lymph from the head region. These will be found, one on each side of the trachea, along with the common carotid artery and internal jugular vein.

Vessels Caudad of the Diaphragm

ARTERIES

(Fig. 6-4, p. 120)

Locate the descending aorta as it passes through the aortic hiatus in the diaphragm. Since the vessels are outside of the peritoneum, this peritoneum must be at least partially dissected away—but *take great care* that nerves and nerve ganglia are not destroyed. Locate the arteries as indicated.

Visceral Branches of the Abdominal Aorta

Celiac trunk. An unpaired artery given off immediately caudad of the diaphragm. Gives branches that supply the stomach, liver, pancreas, duodenum, and spleen. They are the following:

Left gastric. To the lesser curvature of the stomach.

Hepatic. To the liver and other viscera via its branches. The **gastro-duodenal** branch gives a **pyloric artery** to the lesser curvature side of the pylorus, a **right gastroepiploic artery** to the greater curvature side of the pylorus and to the omental bursa, and terminates as the **superior pancreaticoduodenal**, which courses to the duodenum and pancreas. Other branches of the hepatic artery are a **cystic artery** to the gall bladder, and branches to lobes of the liver. (In the human the artery to the lesser curvature side of the pylorus branches directly from the hepatic, and is called the right gastric.)

Splenic. To the greater curvature side of the stomach and to the spleen.

Superior mesenteric. An unpaired artery supplying the pancreas, small intestine, and large intestine as far caudad as the splenic flexure. It gives the following branches:

Inferior pancreaticoduodenal. To the pancreas and duodenum.

Ileocolic. To the ileum and cecum.

Right colic. To the ascending colon. It may be a branch of the ileocolic in some specimens.

Middle colic. To the transverse colon.

Intestinals. Course through the mesentery to the small intestine (jejunum and ileum).

Inferior mesenteric. An unpaired artery to the descending colon and rectum through the following branches:

Left colic. To the descending colon.

Superior hemorrhoidal. To the rectum (and also to the sigmoid colon in the human).

Note the anastomoses between the arteries of the digestive viscera, particularly between the intestinals, and between the arteries supplying the large intestine.

Renal. Paired vessel to the kidneys.

Internal spermatic (in the male), or **ovarian** (in the female). Paired vessel to the reproductive structures.

Adrenolumbar. This paired vessel is a combination visceral and parietal branch to the adrenal glands and body wall. (In the human, the artery to the adrenal gland is separate.)

Parietal Branches of the Abdominal Aorta

Lumbar. Several pairs supplying the body wall.

Iliolumbar. Paired artery to the body wall. (In the human, this is a branch of the internal iliac.)

Fig. 6–4

ARTERIES CAUDAD
OF THE DIAPHRAGM

1. Aorta (descending)
2. Celiac
3. Hepatic
4. Left gastric
5. Splenic
6. Cystic
7. Gastroduodenal
8. Pyloric
9. Right gastroepiploic
10. Superior pancreatico-
    duodenal
11. Superior mesenteric
12. Inferior pancreatico-
    duodenal
13. Middle colic
14. Right colic
15. Ileocolic
16. Intestinals
17. Adrenolumbar
18. Phrenic
19. Renal
20. Lumbars
21. Internal spermatic,
    or ovarian
22. Inferior mesenteric
23. Left colic
24. Superior hemorrhoidal
25. Iliolumbar
26. External iliac
27. Internal iliac
    (hypogastric)
28. Umbilical
29. Superior gluteal
30. Middle hemorrhoidal
31. Inferior gluteal
32. Caudal (middle sacral)
33. Deep femoral
34. Inferior epigastric
35. Branch to urinary bladder
36. Branch to external
    genitalia
37. Lateral femoral circumflex
38. Femoral
39. Branch to muscles
40. Saphenous
41. Popliteal
42. Posterior tibial
43. Anterior tibial

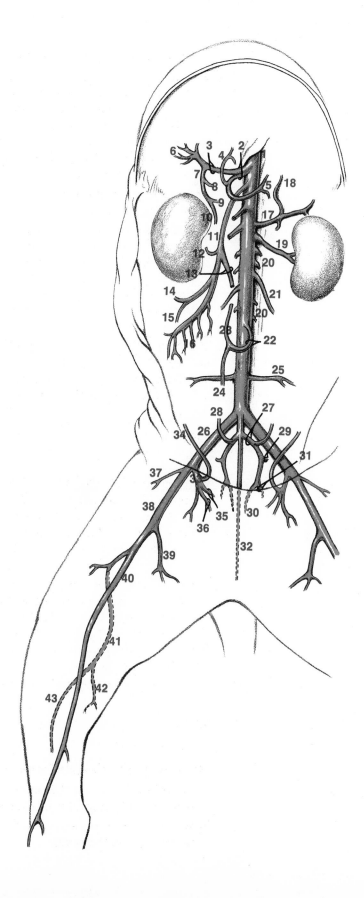

Terminal Branches of the Aorta

External iliac. Paired. It has the following branches:

    Deep femoral. Given off just before the external iliac passes into the thigh to continue as the femoral. A branch of the deep femoral artery, the **inferior epigastric**, supplies the ventral body wall; other branches supply the urinary bladder and external genitalia. (In the human, the deep femoral is a branch of the femoral; the inferior epigastric, of the external iliac.)

Internal iliac (hypogastric). Paired. Its branches go to pelvic viscera and to the gluteal region:

    Umbilical. To the urinary bladder.

    Superior gluteal. To hip muscles.

    Middle hemorrhoidal. To the caudal end of the rectum, urethra, and other tissues in the area. In the female a branch goes to the uterus, and in the male there are branches to the genital structures. (The uterine artery may be a branch of the internal iliac, as it typically is in the human.)

    Inferior gluteal. The terminal portion of the internal iliac. It supplies hip muscles.

Middle sacral. Unpaired. This small artery to the coccygeal region is the terminal portion of the aorta.

In the human the descending aorta gives off the **common iliac arteries** and then terminates as the middle sacral artery. Each common iliac divides into an internal and an external iliac artery.

Major Arteries of the Pelvic Extremity

Femoral. The continuation of the external iliac artery. It gives off the following branches and continues as the popliteal artery:

    Saphenous. On the medial side of the extremity, along with the greater saphenous vein and saphenous nerve.

    Branches to muscles

Popliteal. The continuation of the femoral artery into the popliteal fossa. It divides to give off an anterior and a posterior tibial.

Anterior tibial. Passes ventrad through the interosseous membrane and courses through the leg and into the foot.

Posterior tibial. Terminates in the dorsal crural muscles in the cat, but in the human it continues into the foot. In the human, the posterior tibial gives off a **peroneal** branch.

VEINS

(Fig. 6-5, p. 122)

Major Veins of the Pelvic Extremity

Deep veins. These are the anterior and posterior tibial, popliteal, and femoral (and peroneal in the human).

Superficial veins. You have already observed these veins, but you should review them at this point:

Fig. 6–5
VEINS CAUDAD
OF THE DIAPHRAGM
(OTHER THAN THOSE FROM
THE DIGESTIVE VISCERA)

1. Inferior vena cava
2. Hepatic
3. Adrenolumbar
4. Phrenic
5. Renal
6. Internal spermatic,
   or ovarian
7. Lumbars
8. Iliolumbar
9. Common iliac
10. External iliac
11. Internal iliac (hypogastric)
12. Superior gluteal
13. Middle hemorrhoidal
14. Inferior gluteal
15. Deep femoral
16. Inferior epigastric
17. Tributary from urinary
    bladder
18. Tributary from external
    genitalia
19. Caudal (middle sacral)
20. Lateral femoral circumflex
21. Femoral
22. Tributary from muscles
23. Greater saphenous
24. Popliteal
25. Anterior tibial
26. Posterior tibial

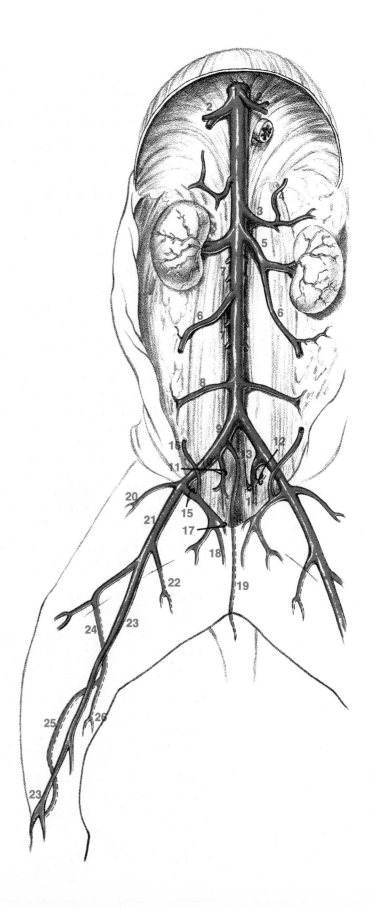

Greater saphenous. Runs along the medial side of the extremity. Joins the femoral vein.

Lesser saphenous. Runs along the dorsal side of the extremity. Joins a tributary of the internal iliac vein.

There are many anastomoses between the superficial veins, and between the superficial and the deep veins.

Veins of the Abdominal Region

Veins of this region, other than those from the digestive viscera and spleen, are the following:

External iliac. A continuation of the femoral vein. It has tributaries paralleling the arteries, and it returns blood distributed by these arteries.

Internal iliac. This vein, with its tributaries, returns blood distributed by the internal iliac artery and its branches. The internal iliac unites with the external iliac to form the common iliac vein.

Common iliac. Formed by the union of the internal and external iliac veins.

Inferior vena cava. Formed by the union of the common iliac veins. Its tributaries are the following:

Paired veins. Correspond to the *paired* branches of the abdominal portion of the aorta and bear the same names: iliolumbar, lumbar, renal, adrenolumbar, and internal spermatic (or ovarian). The left internal spermatic (or ovarian) usually joins the left renal vein rather than the inferior vena cava.

Hepatic veins. These begin in capillaries in the liver (liver sinusoids) and join the inferior vena cava as it passes through the liver. They are variable in number, but there are three in most humans.

The inferior vena cava has no tributaries craniad of the diaphragm, which it passes through before joining the right atrium.

Hepatic Portal System

(Fig. 6-6, p. 124)

The veins of the hepatic portal system will be found coursing through the mesentery and the greater and lesser omenta.

The unpaired **hepatic portal vein** receives blood from all the digestive viscera except the liver, and from the spleen, via veins corresponding to the arteries that distribute blood to these viscera. Find the veins, at least the major ones, using Fig. 6-6 to determine the names and approximate locations.

The hepatic portal vein can be located in the lesser omentum, along with the common bile duct and the hepatic and gastroduodenal arteries. In the cat it is formed by the union of the **gastrosplenic vein** and the **superior mesenteric vein**, the latter having tributaries that correspond to branches of the superior mesenteric artery. In the human the hepatic portal vein is formed by the union of the **splenic vein** and the superior mesenteric vein.

In the cat the **inferior mesenteric vein**, which has tributaries cor-

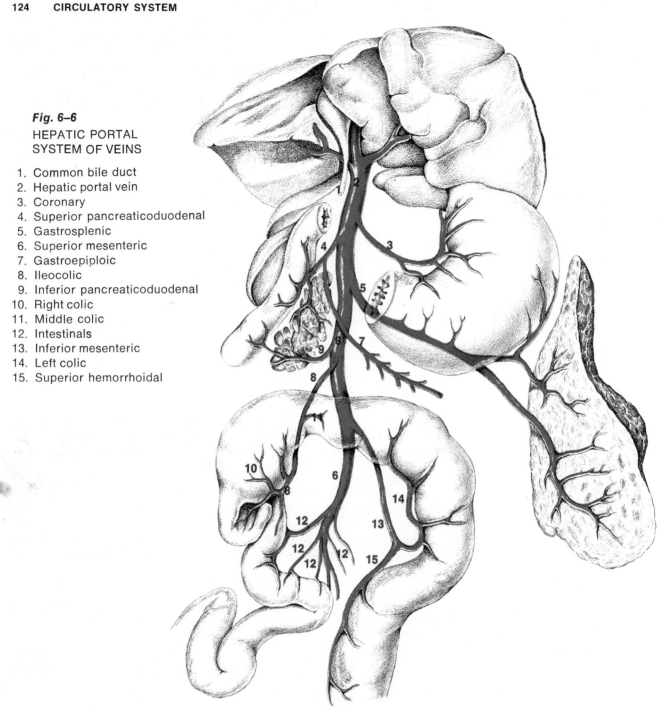

Fig. 6–6

HEPATIC PORTAL
SYSTEM OF VEINS

 1. Common bile duct
 2. Hepatic portal vein
 3. Coronary
 4. Superior pancreaticoduodenal
 5. Gastrosplenic
 6. Superior mesenteric
 7. Gastroepiploic
 8. Ileocolic
 9. Inferior pancreaticoduodenal
10. Right colic
11. Middle colic
12. Intestinals
13. Inferior mesenteric
14. Left colic
15. Superior hemorrhoidal

responding to branches of the inferior mesenteric artery, joins the superior mesenteric. In the human it joins the splenic vein.

The hepatic portal vein divides at the liver to give branches to each lobe that terminate in wide capillaries, called **liver sinusoids**, within the liver substance. The sinusoids are drained by the hepatic veins, which join the inferior vena cava. Blood carried to the liver by the hepatic artery passes through these same sinusoids.

ENDOCRINE GLANDS AND RESPIRATORY AND DIGESTIVE STRUCTURES OF THE HEAD REGION

Endocrine Glands

THYROID GLAND
(Fig. 4-3, p. 98)

The thyroid has two lobes, one on each side of the trachea just caudad of the larynx. The lobes are connected across the ventral surface of the trachea by a narrow band of glandular tissue called the **isthmus**. The isthmus is very small in the cat.

PARATHYROID GLANDS

There are two pairs of parathyroid glands, one cranial and one caudal, lying against the dorsal surface of the lobes of the thyroid gland. These are too small for observation.

THYMUS GLAND
(Fig. 4-3, p. 98)

This gland, which you have already removed, is present in varying degrees in the adult. It lies ventral to the trachea and may extend as far caudad as the heart.

PANCREAS
(Fig. 4-4, p. 102)

You have already studied the pancreas (Chapter 4). The **islets of Langerhans**, which compose the endocrine portion, cannot be seen without the aid of a microscope.

ADRENAL GLANDS
(Figs. 5-1 and 5-2, pp. 107 and 108)

The adrenal glands (or suprarenals) are small oval bodies, one craniad and somewhat mediad of each kidney. (In the human, the adrenals are situated, like caps, immediately over the cranial extremity of each kidney.) You observed these at the time you studied the urogenital system (Chapter 5).

REPRODUCTIVE GLANDS
(Figs. 5-1 and 5-2, pp. 107 and 108)

Certain cells of ovaries and testes are endocrine in function.

PITUITARY GLAND
(Figs. 4-1 and 8-4, pp. 96 and 132)

To locate the pituitary gland (or hypophysis), make a sagittal section of the head, using a bone saw. (Upon occasion it may be desirable to remove the skull cap and lift the whole brain from the cranial cavity. If it is, you must do so before making the sagittal section. Directions for this procedure are given in the reduced section immediately following this paragraph. After you have removed the brain, section the remainder of the head in the sagittal plane.) Saw through the cranium, and hard and soft palate, and the first two or three cervical vertebrae. *Try to leave the tongue intact.* Pull the sections apart so that the epiglottis and the opening to the larynx will be exposed. The pituitary can be observed projecting from the hypothalamus on the ventral side of the brain, within the sella turcica of the sphenoid bone. If the section is exactly through the midline—a difficult achievement—you will also have a section through the midline of the pituitary gland.

DISSECTING FOR THE WHOLE BRAIN

Cut the muscle attachments from the first two or three neural arches and reflect the muscle. Cut the muscle attachments from the occipital and parietal bones, and reflect the muscle. Using bone shears, remove the neural arches to expose the spinal cord; then remove the dorsal, or superior portion of the occipital bone and a part of the lambdoidal ridge (between the parietal and occipital bones) to determine the position of the cerebellum. From this point, cut around the skull cap and lift it off. Note the dura mater of the brain. Carefully remove pieces of bone until the brain is exposed enough so that it can be lifted from the cranial cavity. A bony ledge that separates the cerebellum from the cerebral hemispheres is fused at its outer edge to the parietal bones, and this connection must be cut. To remove the brain from the cranial cavity, sever the spinal cord caudal to the medulla oblongata, sever the nerve and blood vessel connections, and lift the cord from the floor of the vertebral canal. Continue craniad, carefully working the brain loose from the floor of the cranial cavity, severing nerve and blood vessel connections as you proceed. Try to leave a small portion of the nerves and vessels attached to the brain for identification purposes. It will be difficult, if not impossible, to remove the hypophysis and olfactory bulbs intact.

PINEAL GLAND
(Fig. 8-4, p. 132)

This gland is quite small and was probably destroyed if the saw passed through it. If intact, it can be located at the midline, projecting from the caudal border of the roof of the diencephalon.

Other Studies with the Sagittal Section of the Head

Now that you have made a sagittal section of the head, you should observe the most cranial parts of the respiratory and digestive systems. For a list and descriptions of these, refer to Chapter 4, pages 95–98.

Brain

Remove the half section of brain* from one side of the head. Sever the spinal cord caudad of the medulla, and clip all nerves and blood vessels so that you leave a portion of them attached to the brain, if possible. Identify the structures indicated in the following descriptions.

MENINGES

The meninges are the coverings of the brain and also of the spinal cord. They consist of three layers:

Dura mater. The outer layer. This layer consists of two membranes: an outer periosteal, which is close to the inner surface of the skull and usually adherent to it, and an inner meningeal. The two membranes are closely applied to each other for the most part, but in some places there are certain venous channels, the dural sinuses, between the layers.

Arachnoid. The middle layer, which is trabecular (cobweb-like) in structure. You may not be able to distinguish this layer.

Pia mater. The inner layer, which closely invests the brain, dipping into the furrows.

In the living animal the spaces between the meningeal layers are filled with cerebrospinal fluid.

DIVISIONS OF THE BRAIN
(Figs. 8-1, 8-3, and 8-4, pp. 130 and 132)

Forebrain (prosencephalon). The forebrain consists of the following structures:

Telencephalon. Includes the **cerebrum** and **olfactory bulbs**. Note the two large **cerebral hemispheres**, which are characterized by convo-

---

*NOTE: If you removed the whole brain (directions in Chapter 7) when you studied the endocrine glands, now study the brain as a whole, and then make a sagittal section of it when you are ready to study the ventricles.

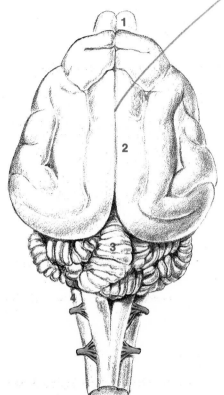

Longitudinal fissure

Fig. 8–1
DORSAL VIEW
OF THE BRAIN

1. Olfactory bulb
2. Cerebral hemisphere
3. Cerebellum
4. Medulla oblongata

lutions (gyri) and furrows (sulci). Note the **corpus callosum**, which is made up of nerve fibers crossing between the hemispheres.

Diencephalon. Composed mainly of the **thalamus** and **hypothalamus.** The hypophysis projects from the hypothalamus at the end of the **infundibular stalk**. Note the **optic chiasma** on the ventral surface, craniad of the hypophysis. The pineal gland (a part of the **epithalamus**) projects from the caudal border of the roof.

Midbrain (mesencephalon). The roof of the midbrain is composed of four bodies called the **corpora quadrigemina** (two **superior colliculi** and two **inferior colliculi**). The floor is composed of **cerebral peduncles**, one on each side, which are made up of ascending and descending nerve fibers.

Hindbrain (rhombencephalon). The hindbrain consists of the following:

Metencephalon. Includes the **cerebellum** dorsally and the **pons** ventrally. Note the many folds of the **cerebellum**. The ventral enlargement on the pons is caused mainly by nerve fibers passing transversely; the dorsal part of the pons is hidden by the cerebellum.

Myelencephalon. This is the **medulla oblongata**. Note its continuity with the spinal cord. The medulla oblongata, pons, and midbrain make up the **brain stem**, which connects the spinal cord with the forebrain.

VENTRICLES
(Fig. 8-4, p. 132)

Within the brain there are a number of spaces that communicate with one another so that there is actually a continuous space, which communicates caudally with the canal of the spinal cord. The space is filled with cerebrospinal fluid in the living animal. The various spaces, from cranial to caudal, are the following:

Lateral ventricles. The first two ventricles are the right and left ventricles of the telencephalon. Each ventricle communicates with ventricle III through an interventricular foramen, the **foramen of Munro**. (The lateral ventricles are not designated by number.)

Ventricle III. In the diencephalon, mainly.

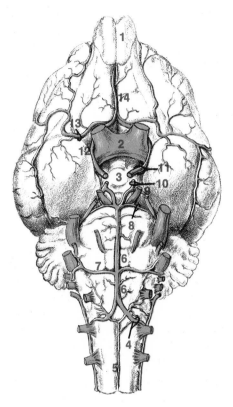

Fig. 8–2
VENTRAL VIEW OF THE BRAIN SHOWING THE ARTERIES

1. Olfactory bulb
2. Optic chiasma
3. Hypophysis
4. Vertebral artery
5. Anterior spinal artery
6. Basilar artery
7. Posterior inferior cerebellar artery
8. Anterior cerebellar artery
9. Posterior cerebral artery
10. Internal carotid artery (cut end)
11. Branch from internal maxillary artery (cut end)
12. Circle of Willis
13. Middle cerebral artery
14. Anterior cerebral artery

Aqueduct (cerebral aqueduct). A narrow canal through the midbrain. It connects the third and fourth ventricles.

Ventricle IV. Located in the hindbrain. It is continuous caudally with the **central canal** of the spinal cord.

BLOOD VESSELS

The rich supply of arteries to the brain can be observed on injected cats and on Fig. 8-2. In general, the arteries are named according to their location and the area they supply. They are continuations, or branches, of the vertebral, internal carotid, and internal maxillary arteries.

The veins of the brain drain into the dural sinuses, which in turn empty into veins that unite to form the internal jugular vein. The vertebral vein also receives blood returning from the brain, as well as from the spinal cord.

Cranial Nerves
(Fig. 8-3, p. 132)

Note any stumps of the cranial nerves that may have remained attached to the brain. These nerves pass through various foramina in the skull and distribute to head structures and some neck structures. (The vagus nerve passes farther caudad.)

The cranial nerves are frequently referred to by number rather than by name. Identify as many of the nerves as possible.

Olfactory (I). This nerve consists of short processes that pass from the olfactory epithelium, in the roof of the nasal cavity, through the cribriform plate of the ethmoid bone to terminate in the **olfactory bulb**. Nerve fibers passing from the olfactory bulb to brain centers make up the **olfactory tract**.

Optic (II). This is the nerve of the retina of the eye. The two optic nerves join in the **optic chiasma** ventral to the diencephalon. Note the **optic tract** caudad of the chiasma.

Oculomotor (III). Emerges from the ventral midbrain. Supplies intrinsic eye muscles (the sphincter muscle of the pupil and the ciliary muscle)

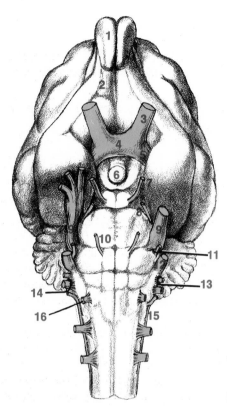

Fig. 8–3

VENTRAL VIEW
OF THE BRAIN
SHOWING THE NERVE
ATTACHMENTS

1. Olfactory bulb
2. Olfactory tract
3. Optic nerve
4. Optic chiasma
5. Optic tract
6. Hypophysis
7. Oculomotor nerve
8. Trochlear nerve
9. Trigeminal nerve
10. Abducens nerve
11. Facial nerve
12. Acoustic nerve
13. Glossopharyngeal nerve
14. Vagus nerve
15. Spinal root of accessory nerve
16. Hypoglossal nerve

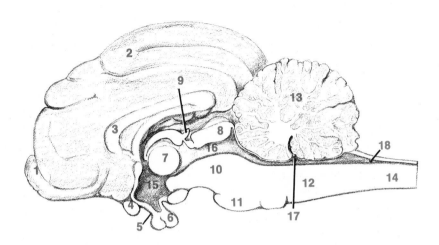

and most of the extrinsic muscles of the eye (the superior, medial, and inferior recti, and the inferior oblique).

Trochlear (IV). Emerges from the roof of the midbrain at its caudal border. Supplies the superior oblique muscle (extrinsic eye muscle).

Trigeminal (V). Emerges from the lateral pons. Provides motor supply for the muscles of mastication, and general sensory fibers for the integument of the head and face, the epithelium of the oral cavity, the anterior two-thirds of the tongue, and for the teeth.

Abducens (VI). Emerges from the ventral pons at the caudal border. Supplies the lateral rectus muscle (extrinsic eye muscle).

Facial (VII). Emerges from the lateral pons. It supplies the cutaneous muscles of facial expression, and also the submandibular, sublingual, and minor salivary glands, the lacrimal gland, and the taste buds of the anterior two-thirds of the tongue; it helps to supply the integument around the auricula of the ear.

Acoustic, or **Auditory (VIII).** Joins the lateral pons at its junction with the medulla. This nerve has two roots, a cochlear and a vestibular. The vestibular portion supplies the semicircular canals and the vestibule of the inner ear (for equilibrium); the cochlear portion supplies the cochlea of the inner ear (for hearing).

Glossopharyngeal (IX). Emerges from the lateral medulla. It supplies muscles and epithelium of the pharynx, the parotid gland, and the epithelium of the posterior third of the tongue (for both taste and general sense).

Vagus (X). Emerges from the lateral medulla. It carries the motor supply for muscles of the larynx and for those of the pharynx not supplied by the glossopharyngeal nerve, and a sensory supply for the epithelium around the epiglottis and in the larynx, and perhaps some sensory supply for the integument around the auricula. It also carries both motor (parasympathetic) and sensory supply to thoracic viscera and to the abdominal viscera as far caudad as the splenic flexure of the colon.

Accessory, or **Spinal Accessory** (XI). Emerges from the lateral medulla. The spinal root supplies muscles of the neck and the cranial component distributes with the vagus nerve.

Hypoglossal (XII). Emerges from the ventral medulla. It supplies muscles of the tongue and some of the hyoid muscles.

The Orbit and its Contents

(Figs. 8-5, 8-6, 8-7, and 8-8, pp. 134–137)

Remove the bony roof of the orbit with bone shears. Note the small superficial muscle, the **levator palpebrae superioris,** that operates the upper eyelid. Cut away the eyelids and free the eyeball from the border of the orbit except at the dorsomedial corner. Note the **lacrimal gland** on the dorsolateral surface of the eyeball. Cut away as much of the malar bone ventrally as is necessary to provide access to the orbit. Carefully pick away the fat and connective tissue, and separate the extrinsic muscles of the eye and blood vessels and nerves.

Various branches (ophthalmics) of the internal maxillary artery, and tributaries (ophthalmics) to the internal maxillary vein and to the deep facial and anterior facial veins will be found in the orbit.

Branches from the **ophthalmic** and **maxillary divisions** of the trigeminal nerve pass through the orbit. One branch from the ophthalmic division (branches of this division are dorsal or medial to the eyeball) supplies the general sensory fibers for the eyeball, but all of the others merely pass through the orbit to supply other structures.

The **infraorbital** branch of the maxillary division of the trigeminal nerve passes across the floor of the orbit, lying on the external pterygoid muscle. (The cat does not have a complete bony orbit.) The infraorbital nerve gives off various branches that reach the epithelium of the oral cavity and the teeth of the upper jaw. (The teeth of the lower jaw are supplied by the **mandibular division** of the trigeminal nerve.) The infraorbital nerve passes through the infraorbital foramen to reach the integument of the face. Two other branches of the maxillary division, the **lacrimal** and **zygomatic,** course mediad of the temporalis muscle in the lateral wall of the orbit. The first passes by the lacrimal gland and leaves the orbit to supply integument over the zygomatic arch; the second, which is more ventral in position, pierces the malar bone to reach the integument of the face.

Locate the extrinsic muscles of the eye, which are given below. All except the inferior oblique arise at the apex of the orbit, which is marked by the optic foramen. (The cat has additional extrinsic muscles, the **retractor oculi,** that lie closer to the eyeball and are partly covered by the other muscles.)

Superior rectus. Insertion on the dorsal surface of the eyeball. It lies deep to the levator palpebrae superioris.

Superior oblique. Passes dorsally along the medial wall of the orbit and inserts on the dorsal surface of the eyeball (dorsolateral in the human). Note the connective tissue "pulley" in the dorsomedial corner, near the rim of the orbit, through which the tendon of insertion passes.

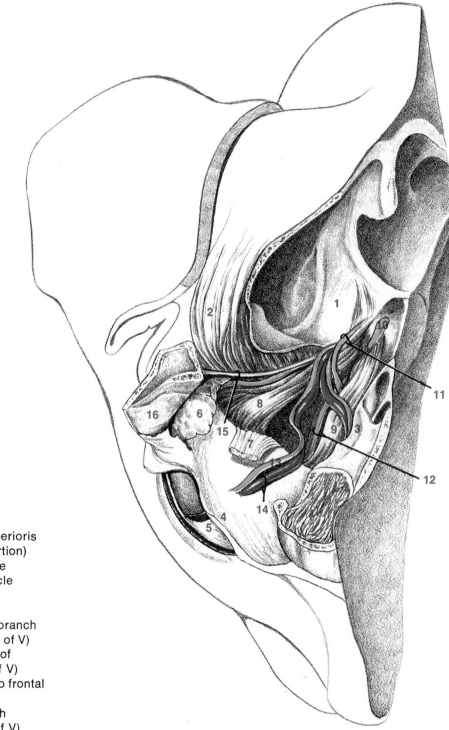

Fig. 8–5
DORSAL VIEW OF
THE CONTENTS
OF THE ORBIT
(RIGHT SIDE)

1. Cerebral fossa
2. Temporalis muscle
3. Medial wall of orbit
4. Base of upper eyelid
5. Nictitating membrane
6. Lacrimal gland
7. Levator palpebrae superioris
   muscle (cut near insertion)
8. Superior rectus muscle
9. Superior oblique muscle
10. Optic nerve (II)
11. Trochlear nerve (IV)
12. Infratrochlear nerve (branch
    of ophthalmic division of V)
13. Frontal nerve (branch of
    ophthalmic division of V)
14. Ophthalmic tributary to frontal
    vein
15. Lacrimal nerve (branch
    of maxillary division of V)
16. Inner surface of malar bone

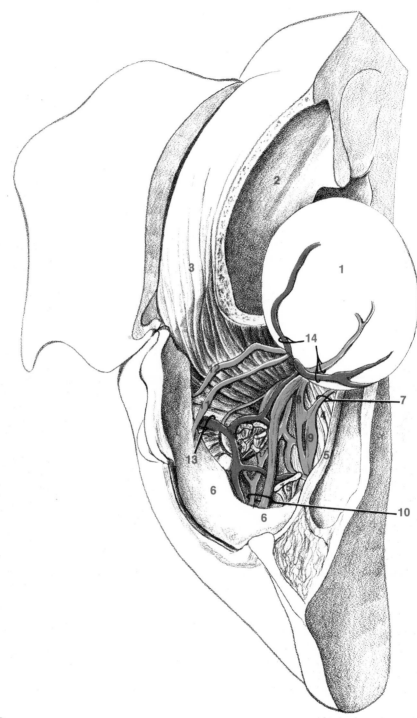

Fig. 8–6

STRUCTURES IN THE
FLOOR OF THE ORBIT
(RIGHT SIDE)

1. Eyeball reflected
2. Cerebral fossa
3. Temporalis muscle
4. External pterygoid muscle
5. Medial wall of orbit
6. Bony portion of floor of orbit
7. Vidian nerve
8. Sphenopalatine nerve
9. Sphenopalatine ganglion
10. Infraorbital nerve (branch
    of maxillary division of V)
    and infraorbital artery
11. Zygomatic nerve (branch
    of maxillary division of V)
12. Lacrimal nerve (branch
    of maxillary division of V)
13. Anastomotic vein between
    superficial temporal
    and deep facial veins
14. Ophthalmic arteries and veins
15. Inferior oblique muscle
    at its origin

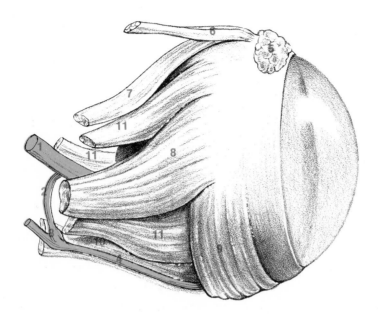

Fig. 8–7

LATERAL VIEW OF
THE RIGHT EYEBALL
SHOWING THE MUSCLE
ATTACHMENTS

1. Optic nerve
2. Short ciliary nerve
3. Ciliary ganglion
4. Branch of oculomotor nerve (III)
   to inferior oblique muscle
5. Lacrimal gland
6. Levator palpebrae superioris
   muscle
7. Superior rectus muscle
8. Lateral rectus muscle
9. Inferior oblique muscle
10. Inferior rectus muscle
11. Retractor oculi muscles

Medial rectus. Insertion on the medial side of the eyeball, ventral to the superior oblique muscle.

Lateral rectus. Insertion on the lateral side of the eyeball.

Inferior oblique. Arises from the medial wall of the orbit and inserts on the ventrolateral surface of the eyeball.

Inferior rectus. Insertion on the ventral surface of the eyeball.

The superior oblique muscle is supplied by cranial nerve IV (the trochlear), the lateral rectus by cranial nerve VI (the abducens), and the other extrinsic muscles by cranial nerve III (the oculomotor). The cranial nerves enter each extrinsic muscle, except the inferior oblique, at a site near the origin of that muscle; the inferior oblique receives a branch from the oculomotor nerve that crosses the ventral surface of the eyeball to reach the muscle.

The **ciliary ganglion** of the oculomotor nerve may be located lying on the inferior rectus muscle. This is a terminal autonomic ganglion from which postganglionic fibers (parasympathetic) pass to two intrinsic eye muscles: the **ciliaris**, which operates the lens, and the **sphincter pupillae.** (The remaining intrinsic muscle, the **dilator pupillae** receives a sympathetic supply.) Another terminal autonomic ganglion, the **sphenopalatine**, lies medial to the infraorbital nerve. Parasympathetic preganglionic fibers from cranial nerve VII (the facial) course through the **Vidian** nerve, which is medial to the infraorbital, to synapse in the sphenopalatine ganglion, from which postganglionic fibers pass to the lacrimal gland and to minor salivary glands in the palate. Sympathetic postganglionic fibers also course in the Vidian nerve and pass to these same glands.

Note the optic nerve connection with the eyeball and its passage through the optic foramen.

Make a sagittal incision through the eyeball (see Fig. 8-8). (It is not necessary to remove the eyeball from the orbit unless your instructor di-

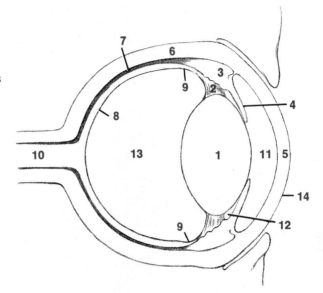

Fig. 8-8

SAGITTAL SECTION
OF THE EYEBALL

1. Crystalline lens
2. Suspensory ligament of lens
3. Ciliary body
4. Iris
5. Cornea
6. Sclera
7. Choroid
8. Retina
9. Ora serrata of retina
10. Optic nerve
11. Anterior chamber
12. Posterior chamber
13. Vitreous body
14. Conjunctiva

rects you to.) Note the tunics of the eyeball: an outer fibrous **sclera**, a middle, highly vascular, thin **choroid** (which will be black or dark brown), and a thin inner **retina** which contains nerve cells for vision. Note the translucent **cornea, crystalline lens,** and **iris.** The aperture in the iris is the **pupil.** The transparent epithelium that covers the exposed surface of the eyeball, and is continuous with epithelium of the eyelids, is the **conjunctiva.** The **ciliary body,** which rings the outer edge of the iris, is attached to the lens by a delicate supensory ligament of radiating fibrils. This band of fibrils is called the **zonula ciliaris,** or **zonule of Zinn,** as well as the **suspensory ligament of the lens.**

Note the chambers of the eye. The **anterior chamber,** between the iris and the cornea, and the **posterior chamber,** bounded by the iris and by the lens and zonula ciliaris, contain a fluid called **aqueous humor** in the living animal. The cavity behind the lens and zonula ciliaris, is filled with the viscous **vitreous body** in the living animal.

Vagus Nerve and Some Other Cranial Nerves

Locate the vagus nerve (Fig. 8-9, p. 138) in the cervical region, where it will be found alongside the cervical portion of the sympathetic trunk, common carotid artery, and internal jugular vein.

Note the large **nodose ganglion** (also called the inferior ganglion, since the vagus has another superior, or jugular ganglion) laterad, and slightly craniad, of the thyroid cartilage. It is laterad of, and closely bound to the superior cervical ganglion of the sympathetic trunk. It may be impossible to separate the two ganglia. Note the **superior laryngeal** branch to the larynx. The **inferior,** or **recurrent laryngeal** branch loops around the subclavian artery on the right, but around the arch of the aorta on the left, and ascends on the surface of the trachea to reach the larynx, where it supplies motor fibers to the muscles. As the vagus nerve courses caudad,

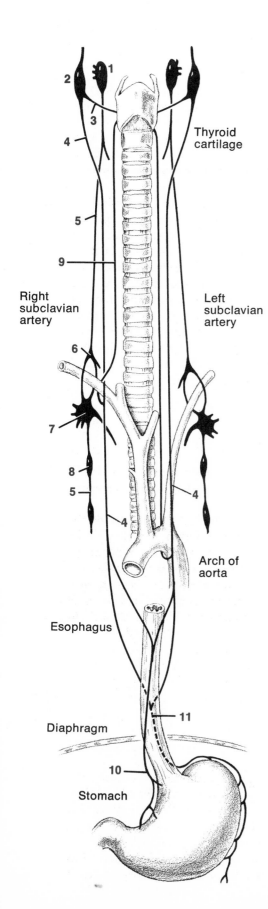

Right
subclavian
artery

Left
subclavian
artery

Thyroid
cartilage

Arch of
aorta

Esophagus

Diaphragm

Stomach

Fig. 8–9

VAGUS NERVE AND SYMPATHETIC TRUNK

1. Superior cervical ganglion
2. Nodose ganglion (inferior ganglion of vagus)
3. Superior laryngeal nerve
4. Vagus nerve
5. Sympathetic trunk
6. Middle cervical ganglion
7. Stellate ganglion
8. Chain ganglion
9. Recurrent laryngeal nerve (inferior laryngeal)
10. Ventral vagal trunk
11. Dorsal vagal trunk

it sends preganglionic (parasympathetic) and sensory fibers to all major thoracic viscera.

Caudad of the bronchus, the vagus divides into ventral and dorsal portions. The ventral divisions from each side unite to form the **ventral vagal trunk**, which courses caudad ventral to the esophagus and passes through the diaphragm to reach the lesser curvature of the stomach, which it supplies. The dorsal divisions from each side course caudad and unite to form the **dorsal vagal trunk**, which continues caudad dorsal to the esophagus and passes through the diaphragm to reach the greater curvature of the stomach, which it supplies. Preganglionic parasympathetic and sensory fibers, primarily from the dorsal trunk, are distributed to major abdominal viscera as far caudad as the splenic flexure of the colon. (Viscera caudad of the splenic flexure receive a parasympathetic supply from sacral spinal nerves.)

There are some other cranial nerves, or branches, which you can easily locate in the head and cervical regions. The hypoglossal nerve accompanies the lingual artery to the tongue. The accessory nerve pierces the cleidomastoid muscle, which it supplies, and sends branches to the sternomastoid and to parts of the trapezius. The glossopharyngeal nerve lies immediately ventral to the skull and passes to the pharynx and to epithelium of the posterior third of the tongue. The ventral and dorsal rami of the facial nerve are easily located. The ventral ramus is superficial to the lower border of the masseter muscle (craniad of the submandibular and sublingual glands), and the dorsal ramus is immediately in front of the external ear, deep to the parotid gland. The ventral ramus sends branches to the superficial muscles around the mouth; the dorsal ramus to those around the ear and eye, and the cheek. Other superficial branches of the facial nerve go to the posterior part of the digastric muscle and to superficial muscles on the dorsal surface of the skull.

Spinal Cord and Spinal Nerves

You will not dissect the spinal cord, but you will study the nerves that are connected with it. The human has 31 pairs of spinal nerves: 8 cervical, 12 thoracic, 5 lumbar, 5 sacral, and 1 caudal or coccygeal. The cat has 38 or 39: 8 cervical, 13 thoracic, 7 lumbar, 3 sacral, and 7 or 8 caudal or coccygeal. A spinal nerve is formed by the union of a **ventral root** (motor) and a **dorsal root** (sensory); on the dorsal root is located a group of sensory nerve cell bodies which form an enlargement called the **dorsal root ganglion,** or **spinal ganglion.** The spinal nerve formed from the two roots soon divides into a **dorsal ramus** and a **ventral ramus.** Dorsal rami supply the deep muscles of the back and neck and other structures of the dorsal body wall. Ventral rami distribute to the superficial muscles of the back and neck, to muscles and other structures of the ventral and lateral body wall, and to the appendages.

Associated with each spinal nerve is a group of nerve cell bodies that form a ganglion. These ganglia, called chain ganglia, are interconnected by ascending and descending neuron processes, so that a continuous trunk, the sympathetic trunk, is formed. Some of these neuron processes are those of postganglionic neurons with cell bodies in the chain ganglia; some are those of preganglionic neurons, with cell bodies in the spinal cord, that will synapse in chain ganglia at levels that are higher or lower than the point of emergence of the processes from the spinal cord; some are those of sensory neurons from viscera. Only thoracic and upper lumbar nerves contribute preganglionic processes to the sympathetic trunk, but all spinal nerves have chain ganglia associated with them, and all contribute postganglionics.

Sympathetic Trunk and Branches
(Figs. 8-9 and 8-10, pp. 138 and 140)

Locate the sympathetic trunk in the cervical region, where it is closely adjacent to the vagus nerve and may be so bound to it by connective tissue that it appears to be a part of the vagus.

Locate the three sympathetic ganglia that are fusions of some of the original separate ganglia. The **superior cervical ganglion** is medial to the nodose ganglion. This is a combination of the original chain ganglia of the first four cervical nerves. Postganglionic processes leave the ganglion and follow the common carotid artery and its branches to the head region. The **middle cervical ganglion,** which is just craniad of the subclavian artery, is very small. This ganglion represents a fusion of the original chain ganglia of the fifth and sixth cervical nerves. Caudad of the middle cervical ganglion, the sympathetic trunk splits and forms a loop around the subclavian artery. It then joins the **stellate ganglion,** which lies laterad of the vertebral bodies between the first and second ribs. The stellate ganglion represents fused chain ganglia of the seventh and eighth cervical nerves and the first four thoracic nerves. (In the human the ganglia of the seventh and eighth cervical nerves fuse to form an inferior cervical ganglion, which may or may not fuse with one or more thoracic chain ganglia to form a stellate ganglion.)

All three of these sympathetic ganglia give off nerves to the heart: the superior cervical ganglion gives off the **superior cardiac nerve;** the middle cervical ganglion, the **middle cardiac nerve;** the stellate ganglion, the **inferior cardiac nerve.** They also give off fibers to other thoracic viscera.

Caudad of the stellate ganglion the sympathetic trunk lies laterad of the bodies of the vertebrae, and small chain ganglia can be observed at intervals. Before passing through the diaphragm, the sympathetic trunk

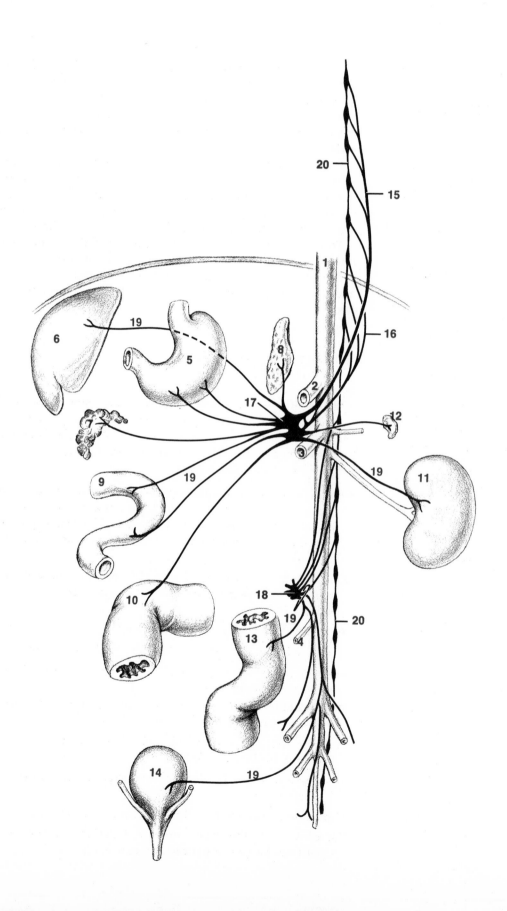

Fig. 8–10

SYMPATHETIC TRUNK
AND BRANCHES TO
THE ABDOMINAL VISCERA

1. Aorta
2. Celiac artery
3. Superior mesenteric artery
4. Inferior mesenteric artery
5. Stomach
6. Liver
7. Pancreas
8. Spleen
9. Small intestine
10. Ascending and transverse colon
11. Kidney
12. Adrenal gland
13. Descending colon and rectum
14. Urinary bladder
15. Greater splanchnic nerve
16. Lesser splanchnic nerve
17. Celiac and superior mesenteric ganglia
18. Inferior mesenteric ganglion
19. Postganglionic fibers
20. Sympathetic trunk

gives off the **greater** and **lesser splanchnic nerves** which contain preganglionic processes that synapse in collateral, or intermediate, ganglia. These ganglia are the **celiac**, near the base of the celiac artery, and the **superior mesenteric**, near the base of the superior mesenteric artery. Postganglionic processes leave the ganglia to follow the arteries and their branches to the viscera.

Preganglionic fibers are given off from the lumbar portion of the sympathetic trunk and course caudad to synapse in the **inferior mesenteric ganglion**, a collateral ganglion near the base of the inferior mesenteric artery. Postganglionics then follow the artery and its branches to the viscera. Postganglionics also follow the aorta and its terminal branches.

Postganglionic fibers are given off by all of the chain ganglia to supply minor viscera and integumentary glands in the area.

Note that the sympathetic trunk, as it courses caudad, swings mediad and occupies a position ventral to the vertebral bodies before its termination in the sacral region.

Brachial Plexus
(Fig. 8-11, p. 142)

The brachial plexus is formed by the ventral rami of the last four cervical nerves and the first thoracic nerve. The peripheral nerves arising from this plexus, which is located in the axilla, supply the pectoral appendage. Some of the nerves you have observed previously.

Locate the following nerves:

Phrenic. Goes to the diaphragm. Near its origin it accompanies the thyrocervical trunk. (The phrenic nerve is not actually a part of the brachial plexus. In the human it arises from a small cervical plexus; it is mainly from the fourth cervical nerve, but receives contributions from the third and fifth.)

Suprascapular. Passes over the cranial border of the scapula. Follows the transverse scapular artery to the supraspinatus and infraspinatus muscles.

First subscapular. Goes to the subscapularis muscle.

Second subscapular. Goes to the subscapularis and teres major muscles.

Axillary. Follows the posterior humeral circumflex artery to the teres minor and deltoid muscles.

Musculocutaneous. Goes to the ventral muscles of the arm, and gives a cutaneous branch to the forearm.

Posterior thoracic, or **long thoracic.** Goes to the serratus anterior and levator scapulae muscles.

First anterior thoracic. Goes to the pectorales. Accompanies the anterior thoracic artery.

Third subscapular. Follows the thoracodorsal artery to the latissimus dorsi.

Radial. Goes to the dorsal muscles of the arm and forearm. Supplies a cutaneous branch to the forearm and hand.

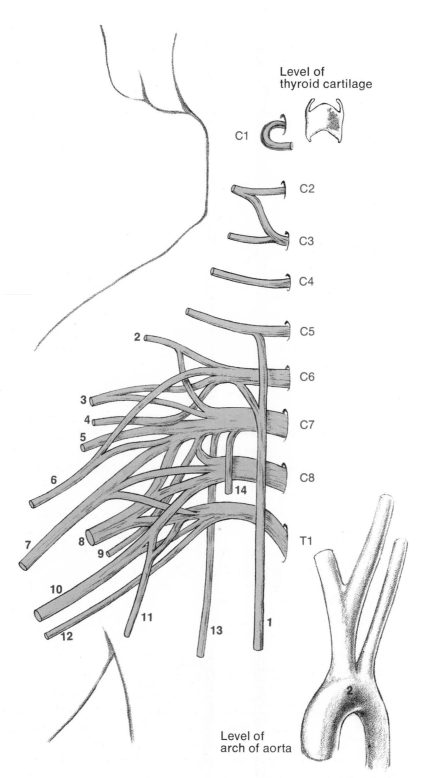

Level of
thyroid cartilage

C1

C2

C3

C4

C5

C6

C7

C8

T1

Level of
arch of aorta

Fig. 8–11
NERVES OF
THE BRACHIAL PLEXUS

1. Phrenic nerve
2. Suprascapular
3. First subscapular
4. Axillary
5. Second subscapular
6. Musculocutaneous
7. Median
8. Radial
9. Third subscapular
10. Ulnar
11. Second anterior thoracic
12. Medial cutaneous
13. Posterior thoracic (long thoracic)
14. First anterior thoracic

Median. Supplies the ventral muscles of the forearm, except for the flexor carpi ulnaris and the ulnar head of the flexor digitorum profundus; supplies also some of the intrinsic hand muscles. In the cat this nerve accompanies the brachial and radial blood vessels.

Ulnar. Passes dorsal to the medial epicondyle of the humerus and supplies the flexor carpi ulnaris and the ulnar head of the flexor digitorum profundus. Also supplies some of the intrinsic hand muscles.

Second anterior thoracic. Supplies the latissimus dorsi and the pectorales. Accompanies the long thoracic artery.

Medial cutaneous. Passes distad on the medial side of the arm to integument on the medial side of the forearm.

Lumbosacral Plexus
(Fig. 8-12, p. 144)

The lumbosacral plexus is formed by ventral rami of lumbar and sacral spinal nerves. The nerves that contribute to the plexus vary slightly between the cat and the human. The peripheral nerves arising from the plexus, which is located in the lower abdominal and pelvic regions, supply the integument and muscles of the pelvic region and the pelvic appendage. Some of the nerves you have observed previously.

Locate as many of the following nerves as possible:

Genitofemoral. Supplies integument of the external genitalia, and the medial side of the thigh and adjacent abdominal body wall. This nerve has two branches, a medial and a lateral. The medial branch courses along the medial side of the iliopsoas and then accompanies the external iliac artery and its deep femoral branch. The lateral branch pierces the psoas minor and courses caudad ventral to the muscle, crosses the iliolumbar blood vessels, and proceeds caudad to its distribution.

Lateral femoral cutaneous. Supplies the integument covering the lateral surface of the thigh and hip region. This nerve will be found close to the iliolumbar blood vessels.

Femoral. Emerges from the psoas major and distributes to the ventral femoral muscles. It gives off a cutaneous branch, the **saphenous nerve**, which accompanies the greater saphenous vein and saphenous artery, to the leg and foot.

Obturator. Passes through the obturator foramen and distributes to the medial femoral muscles and the obturator externus.

Just distal to the origin of the obturator nerve, and dorsal to the iliac blood vessels, you will observe a large nerve cord. This is the **lumbosacral cord**, or **trunk**; it is composed of portions of ventral rami of the sixth and seventh lumbar nerves, which join the ventral rami of sacral nerves. (Since the human has only five pairs of lumbar nerves, the lumbosacral cord is composed of a branch of the ventral ramus of the fourth lumbar and the ventral ramus of the fifth lumbar nerve.)

Superior gluteal. Passes out of the pelvic region with the superior gluteal blood vessels. Supplies most of the muscles of the dorsal hip region.

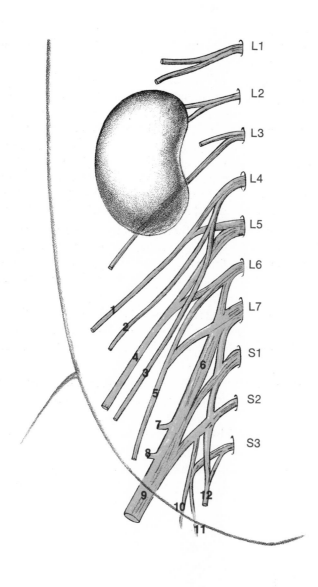

Fig. 8–12
NERVES OF THE
LUMBOSACRAL PLEXUS

1. Lateral femoral cutaneous
2. Genitofemoral, lateral branch
3. Genitofemoral, medial branch
4. Femoral
5. Obturator
6. Lumbosacral cord
7. Superior gluteal
8. Inferior gluteal
9. Sciatic
10. Posterior femoral cutaneous
11. Pudendal
12. Inferior hemorrhoidal

Inferior gluteal. Can be located dorsally, deep to the caudofemoralis and gluteus maximus, both of which it supplies.

Sciatic. A very large nerve, which you have already observed deep to the caudofemoralis, gluteus maximus, and biceps femoris. It supplies, through its branches, all of the dorsal femoral muscles and all of the crural muscles and intrinsic foot muscles. Its terminal branches are the **common peroneal** and **tibial nerves,** which you have observed. The common peroneal nerve divides, in turn, into a deep branch that supplies the ventral crural muscles and a superficial branch that supplies the lateral crural muscles; these branches continue into the foot. The tibial nerve supplies the dorsal crural muscles and continues into the foot.

Pudendal. Supplies muscles and other tissues in the area of the external genitalia. It also gives off branches to the caudal end of the rectum and anal area. The nerve can be located laterad of the levator ani muscle, which is a flat sheet in the pelvic floor that can be detached and reflected.

Posterior femoral cutaneous. Accompanies the inferior gluteal blood vessels and extends onto the thigh, where it can be located in the connective tissue on the lateral surface of the biceps femoris. It sends branches also to the anal area.

Inferior hemorrhoidal. Follows the middle hemorrhoidal artery to the urethra, urinary bladder, and caudal end of the rectum.

Index